MIX
Papier aus verantwortungsvollen Quellen
Paper from responsible sources
FSC® C105338

Azees Maria

Efficient Anonymous Authentication and Key Management Techniques for Vehicular Ad-hoc Networks

Anchor Academic Publishing

Maria, Azees: Efficient Anonymous Authentication and Key Management Techniques for Vehicular Ad-hoc Networks, Hamburg, Anchor Academic Publishing 2017

Buch-ISBN: 978-3-96067-180-0
PDF-eBook-ISBN: 978-3-96067-680-5
Druck/Herstellung: Anchor Academic Publishing, Hamburg, 2017

Bibliografische Information der Deutschen Nationalbibliothek:
Die Deutsche Nationalbibliothek verzeichnet diese Publikation in der Deutschen Nationalbibliografie; detaillierte bibliografische Daten sind im Internet über http://dnb.d-nb.de abrufbar.

Bibliographical Information of the German National Library:
The German National Library lists this publication in the German National Bibliography. Detailed bibliographic data can be found at: http://dnb.d-nb.de

All rights reserved. This publication may not be reproduced, stored in a retrieval system or transmitted, in any form or by any means, electronic, mechanical, photocopying, recording or otherwise, without the prior permission of the publishers.

Das Werk einschließlich aller seiner Teile ist urheberrechtlich geschützt. Jede Verwertung außerhalb der Grenzen des Urheberrechtsgesetzes ist ohne Zustimmung des Verlages unzulässig und strafbar. Dies gilt insbesondere für Vervielfältigungen, Übersetzungen, Mikroverfilmungen und die Einspeicherung und Bearbeitung in elektronischen Systemen.

Die Wiedergabe von Gebrauchsnamen, Handelsnamen, Warenbezeichnungen usw. in diesem Werk berechtigt auch ohne besondere Kennzeichnung nicht zu der Annahme, dass solche Namen im Sinne der Warenzeichen- und Markenschutz-Gesetzgebung als frei zu betrachten wären und daher von jedermann benutzt werden dürften.

Die Informationen in diesem Werk wurden mit Sorgfalt erarbeitet. Dennoch können Fehler nicht vollständig ausgeschlossen werden und die Diplomica Verlag GmbH, die Autoren oder Übersetzer übernehmen keine juristische Verantwortung oder irgendeine Haftung für evtl. verbliebene fehlerhafte Angaben und deren Folgen.

Alle Rechte vorbehalten

© Anchor Academic Publishing, Imprint der Diplomica Verlag GmbH
Hermannstal 119k, 22119 Hamburg
http://www.diplomica-verlag.de, Hamburg 2017
Printed in Germany

ABSTRACT

The Vehicular ad hoc network (VANET) is an important communication paradigm in modern-day transport system for exchanging live messages regarding traffic congestion, weather conditions, road conditions, and targeted location-based advertisements to improve the driving comfort. In such environments, authentication and privacy are two important challenges needed to be addressed.

There are many existing works to provide authentication and privacy in VANETs. However, most of the existing authentication schemes are suffered from high computational cost during authentication and high communicational cost during secure key distribution to a group of vehicles. Moreover, in many existing schemes, there is no conditional tracking mechanism is available to revoke the misbehaving vehicles from the VANET system. In order to overcome these issues, four new approaches have been developed in this research work.

Firstly, a dual authentication scheme is developed to provide a high level of security in the vehicle side to effectively prevent the unauthorized vehicles entering into the VANET. Moreover, a dual group key management scheme is developed to efficiently distribute a group key to a group of users and to update such group keys during the users' join and leave operations. The major advantage of the proposed dual key management is that adding/revoking users in the VANET group can be performed in a computationally efficient manner by updating a small amount of information. The results of the proposed dual authentication and key management scheme are computationally efficient compared with all other existing schemes discussed in literature, and the results are promising.

Secondly, in order to preserve the privacy of vehicle users, a computationally efficient privacy preserving anonymous authentication scheme (CPAV) is developed to anonymously authenticate the vehicle users based on the use of anonymous certificates and signatures. Even though there were many existing schemes to provide anonymous authentication based on anonymous certificates and signatures in VANETs, the existing schemes suffer from high computation cost in the certificate revocation list (CRL) checking process and in the certificate and the signature verification process. Therefore, a computationally efficient anonymous mutual authentication mechanism is proposed in this research work to preserve the privacy of the vehicle users and to guarantee the integrity of the transmitted messages. Moreover, a conditional tracking mechanism is introduced to trace the real identity of vehicles and revoke them from VANET in the case of dispute.

Thirdly, an efficient anonymous authentication scheme to preserve the privacy of RSUs is proposed in this research work. In this research work, each authenticated vehicle is required to authenticate the RSUs in an anonymous manner, before communicating with it. Because, each RSU provides the location based safety information (LBSI) to all authenticated vehicles when they are entered into its region. By doing this, each RSU provides the knowledge to vehicle users about the obstacles within its coverage area.

Finally, a computationally efficient group key distribution (CEKD) scheme for secure group communication is proposed in this research work based on bilinear pairing. In VANETs, secure and reliable group communication is an energetic area of research. Today, the most important research challenge is an efficient group key distribution for a secure group communication. Even though there are many group key distribution protocols, they have the security and performance weakness. The proposed CEKD scheme provides better performance in comparison with most of the previously proposed key distribution schemes in terms of computation cost and hence it is suitable for secure group communication in VANETs.

ACKNOWLEGDEMENT

I, with great pleasure would like to express my heartfelt thanks to my esteemed research supervisor **Dr. P. Vijayakumar**, Assistant Professor, University College of Engineering Tindivanam, Tindivanam, for his persistent help, continued drive and timely motivation which has made this work possible. His illuminating comments and genuine suggestions enabled me to carry out this work fruitfully.

I am very much grateful to **Dr.D.Loganathan**, Professor, Department of Computer Science and Engineering, Pondicherry Engineering College, Puducherry, and to **Dr.K.Kulothungan**, Assistant Professor, Department of Information Sciences and Technology, CEG Campus, Anna University, Chennai, for acting as the doctoral committee members and to provide their valuable suggestions and encouragements throughout the period of my research.

I sincerely express my great sense of gratitude to **Dr. L. Jegatha Deborah**, Assistant Professor and Head i/c, Department of Computer Science and Engineering, University College of Engineering Tindivanam, Tindivanam for the support rendered to me at all the stages of my research.

The whole task of acknowledging seems to be incomplete if I don't owe my indebtedness and gratitude to my parents **Mr. V. Maria John Francis** and **Mrs. G. Martinammal** and my brother **Mr. M. Abeens** for their invaluable moral support at every stage of my progress in this research work. Not to mention, my family is the greatest strength behind all my endeavors. Above all, I thank **God, the Almighty** for having blessed me with all physical and mental strength in executing my will successfully.

AZEES M

TABLE OF CONTENTS

CHAPTER NO.	TITLE	PAGE NO.
	ABSTRACT	i
	LIST OF TABLES	x
	LIST OF FIGURES	xi
	LIST OF SYMBOLS AND ABBREVIATIONS	xii
1	INTRODUCTION	1
	1.1 VANET OVERVIEW	2
	1.1.1 VANET System Model	2
	1.1.2 Dedicated Short Range Communication (DSRC)	5
	1.1.3 VANET Characteristics	6
	1.2 SECURITY ISSUES IN VANET	7
	1.3 PROPOSED WORKS	9
	1.4 OBJECTIVES OF THE RESEARCH WORK	10
	1.5 ASSUMPTIONS	11
	1.6 ORGANIZATION OF THE THESIS	11
2	LITERATURE SURVEY	13
	2.1 INTRODUCTION	13
	2.2 SECURITY SERVICES OF VANETS	13
	2.3 AVAILABILITY IN VANETS	14
	2.3.1 Threats and Attacks on Availability	15
	2.3.2 Works on Availability	17
	2.4 CONFIDENTIALITY IN VANETS	18
	2.4.1 Threats and Attacks on Confidentiality	19

CHAPTER NO.		TITLE	PAGE NO.
	2.4.2	Works on Confidentiality	20
2.5	AUTHENTICATION IN VANETS		21
	2.5.1	Threats and Attacks on Authentication	21
	2.5.2	Requirements for Authentication	23
	2.5.3	Works on Authentication with Privacy Preservation	24
	2.5.4	Computational Cost for Various Authentication Schemes	32
2.6	DATA INTEGRITY IN VANETS		34
	2.6.1	Threats and Attacks on Data Integrity	34
	2.6.2	Works on Data Integrity	36
2.7	NON-REPUDIATION IN VANETS		37
	2.7.1	Attack on Non-repudiation	37
	2.7.2	Works on Non-repudiation	37
2.8	COUNTER MEASURES ON VARIOUS SECURITY ATTACKS		38
2.9	WORKS ON KEY MANAGEMENT		41
2.10	LITERATURE SURVEY GAPS		42
2.11	PROPOSED WORK		43
2.12	CONCLUSIONS		43
3	**SYSTEM ARCHITECTURE**		**45**
4	**DUAL AUTHENTICATION AND DUAL KEY MANAGEMENT FOR GROUP COMMUNICATION**		**48**
	4.1	INTRODUCTION	48
	4.2	PROPOSED DUAL AUTHENTICATION TECHNIQUE	50

CHAPTER NO.			TITLE	PAGE NO.
		4.2.1	Registration through Offline Mode	52
		4.2.2	Vehicle's Authentication Process	53
		4.2.3	Trusted Authority's Authentication Process and the Provision of Authentication Code (AC)	54
	4.3		PROPOSED DUAL KEY MANAGEMENT FOR GROUP COMMUNICATION	58
		4.3.1	TA Initial Set up	60
		4.3.2	Group Key Computation	61
		4.3.3	Secure Data Transmission in VANETs	63
		4.3.4	Key Updating	65
	4.4		SECURITY ANALYSIS	68
		4.4.1	Resistance to Replay Attack	68
		4.4.2	Masquerade and Sybil Attacks	68
		4.4.3	Message Tampering /Fabrication/ Alteration Attack	69
		4.4.4	Backward Secrecy	69
		4.4.5	Forward Secrecy	70
		4.4.6	Collusion Attack	71
	4.5		PERFORMANCE ANAYSIS	72
	4.6		CONCLUSIONS	76
5			CPAV: COMPUTATIONALLY EFFICIENT PRIVACY PRESERVING ANONYMOUS AUTHENTICATION FOR A VEHICLE USER IN VANETs	78
	5.1		INTRODUCTION	78
	5.2		SECURITY REQUIREMENTS	78

CHAPTER NO.	TITLE	PAGE NO.

	5.3	BILINEAR PAIRING		79
	5.4	PROPOSED CPAV SCHEME		80
		5.4.1	System Initialization	80
		5.4.2	Registration	81
		5.4.3	Secure Activation Key Distribution	81
		5.4.4	CPAV Secure Anonymous Mutual Authentication	82
	5.5	SECURITY ANALYSIS		85
		5.5.1	Message Integrity and Source Authentication	85
		5.5.2	Conditional Privacy Preservation	86
		5.5.3	Anonymity	86
	5.6	PERFORMANCE ANAYSIS		87
	5.7	CONCLUSIONS		91
6	EFFICIENT ANONYMOUS AUTHENTICATION OF AN RSU			92
	6.1	INTRODUCTION		92
	6.2	ANONYMOUS AUTHENTICATION		93
		6.2.1	System Initialization	93
		6.2.2	Anonymous Authentication of an RSU	94
	6.3	SECURITY ANALYSIS		98
	6.4	PERFORMANCE ANALYSIS		99
		6.4.1	RSU Serving Capability	100
	6.5	CONCLUSIONS		102

CHAPTER NO.	TITLE	PAGE NO.

7 CEKD: COMPUTATIONALLY EFFICIENT KEY DISTRIBUTION   103
    7.1 INTRODUCTION   103
    7.2 CEKD SCHEME   104
        7.2.1 System Initialization   104
        7.2.2 VANET License Issuing   104
        7.2.3 CEKD Scheme   105
    7.3 SECURITY ANALYSIS   107
    7.4 PERFORMANCE ANALYSIS   108
    7.5 CONCLUSIONS   110

8 CONCLUSIONS AND FUTURE WORKS   111
    8.1 DUAL AUTHENTICATION AND DUAL KEY MANAGEMENT FOR GROUP COMMUNICATION   111
    8.2 CPAV: COMPUTATIONALLY EFFICIENT PRIVACY PRESERVING ANONYMOUS AUTHENTICATION   112
    8.3 EFFICIENT ANONYMOUS AUTHENTICATION OF AN RSU   112
    8.4 CEKD: COMPUTATIONALLY EFFICIENT KEY DISTRIBUTION   113
    8.5 FUTURE WORKS   113

    REFERENCES   114

LIST OF TABLES

TABLE NO.	TITLE	PAGE NO.
2.1	Comparison of authentication techniques	31
2.2	Certificate and signature verification cost of various schemes	33
2.3	Security problems and their countermeasures	39
4.1	Computation, storage and communication complexities of various schemes	74
5.1	Certificate and signature verification cost of various schemes	88
5.2	Comparisons of total computational delay of various schemes	90
6.1	Location based safety information (LBSI)	96
7.1	Computational cost of various key distribution schemes	109

LIST OF FIGURES

FIGURE NO.	TITLE	PAGE NO.
1.1	VANET system model	4
1.2	Channel layout for DSRC	5
3.1	System architecture	45
4.1	Authentication in vehicle and the TA	52
4.2	Secure data communication in VANET	65
4.3	Group key computation time at TA side	75
4.4	PUs key recovery time in the VANET	76
5.1	Comparison of certificate and signature Verification time	89
5.2	Comparison of total computational delay	91
6.1	RSU serving capability of EAAP scheme for various vehicle density d and various average speed of vehicles v, when RSU range r = 300 m	101
7.1	Computational cost for key distribution of various schemes	109

LIST OF SYMBOLS AND ABBREVIATIONS

s	-	Activation key
AC	-	Authentication code
e	-	Bilinear map
BLS	-	Boneh-Lynn-Shacham
CRL	-	Certificate revocation list
CAS	-	Certificateless aggregate signatures
CRGK	-	Chinese remainder group key
CRT	-	Chinese remainder theorem
CPAS	-	Conditional privacy-preserving authentication scheme
CMAP	-	Cooperative message authentication protocol
D	-	Decryption
DSRC	-	Dedicated short range communication
DoS	-	Denial of service
DRA	-	Disaster relief authority
ECPP	-	Efficient conditional privacy preservation protocol
EPAS	-	Efficient privacy-preserving authentication scheme
EGKM	-	Elgamal group key management
ECC	-	Elliptical curve cryptosystem
E	-	Encryption
EDR	-	Event data recorder
FRGK	-	Fast-Chinese remainder group key
FCC	-	Federal communications commission
GPS	-	Global positioning system
GSIS	-	Group signature and ID-based signature
GBS	-	Group based signature
HC	-	Hash code
HMAC	-	Hash message authentication code

IBV	-	ID-based batch verification
ID_{TA}	-	Identitity of the TA
ID	-	Identity
ID_{RSU}	-	Identity of the RSU
ID_v	-	Identity of the vehicle
ITS	-	Intelligent transport system
KPSD	-	Key-insulated pseudonym self-delegation
LBSI	-	Location based safety information
LBS	-	Location based services
MAA	-	Message authentication acceleration
MAC	-	Message authentication code
Z_q^*	-	Multiplicative group of order q
NTRU	-	Number theory research unit
OBU	-	On board unit
OFDM	-	Orthogonal frequency division multiplexing
PBC	-	Pairing-based cryptography
pw	-	Password
PU	-	Primary user
PUSK	-	Primary user secret key
PACP	-	Pseudonymous authentication based conditional privacy
PBC	-	Public key cryptography
PKI	-	Public key infrastructure
REK	-	Re-encryption key
RSA	-	Rivest-Shamir-Adleman
RSU	-	Road side unit
RSK	-	RSU secret key
SU	-	Secondary user
SUSK	-	Secondary user secret key
SM	-	Secret message
SHA	-	Secure hash algorithm

SLA	-	Service level agreement
σ	-	Signature
STAR	-	Social tier assisted packet
TPD	-	Tamper proof device
TS	-	Time stamp
TCC	-	Total computational cost
TA	-	Trusted Authority
UU	-	Unauthorized user
VGKM	-	VANET group key management
VSK	-	Vehicle secret key
V2R	-	Vehicle to RSU
V2V	-	Vehicle to vehicle
VANET	-	Vehicular ad-hoc network
VSN	-	Vehicular sensor networks
WAVE	-	Wireless access of vehicular environments
ZBF	-	Zone-based forwarding

CHAPTER 1

INTRODUCTION

Today, many people are using vehicles to travel from one place to another place. The increase in the number of vehicles in the road transport system leads to traffic jams and road fatalities. Traffic jams and road fatalities can be reduced by providing proper information about the road conditions and its surrounding environment to vehicle drivers in a secure way. The increase in critical driving problems lead to road accidents and traffic congestions. For solving these kind of critical driving problems, the vehicles are integrated with communication technologies to exchange information among vehicles and between vehicles and road side units (RSUs). In order to share the associated information about critical driving situations, vehicular ad-hoc networks (VANETs) provide two types of communications, namely, vehicle-to-vehicle (V2V) communication and vehicle to RSU (V2R) communication (Blum & Eskandarian 2004). In V2V communication, vehicles communicate directly with other vehicles to exchange the information. In V2R communication, vehicles communicate directly with RSUs which are fixed aside the roads. Dedicated short range communication (DSRC) radio (Jiang & Delgrossi 2008) is used for V2V and V2R communications in VANETs.

The information shared in VANETs is categorised into two types, namely safety information and non-safety information. In these two types of information, the safety information (e.g. curve speed warning, pedestrian crossing warning) is the primary information to be shared for informing the drivers about expected dangers in order to avoid accidents and traffic jams.

The aim of providing the safety information is to protect the safety of life, health, or property (Robinson 2007). The non-safety information is to enhance the comfort of the drivers and passengers by providing value-added services such as nearest hospital, nearest petrol station and so on (Jakubiak & Koucheryavy 2008). However, safety information have higher priority over the non-safety information. Even though VANET offers many facilities, the adversaries target the wireless medium used in VANETs to make it vulnerable to various kinds of attacks such as interference, eavesdropping, jamming and so on (Dhamgaye & Chavhan 2013). Even though there are several attacks existing for breaking the security of VANET communication, many researchers have developed different approaches (Lin et al. (2007), Hao et al. (2011), Zhang et al. (2008c) and Zhang et al. (2008b)) to improve the security of VANET communications. Basically, the design of VANET security should promise the security services such as availability, confidentiality, authentication, integrity and non-repudiation to guard the VANET network from attackers (Raya & Hubaux 2007). In addition to the basic security services, privacy preservation is one of the major challenges in VANET. In privacy preservation, the user's real identity (ID) and user's location information have to be protected from unwanted users. However, the user's personal information can be revealed by a trusted authority (TA) during crimes or dispute.

1.1 VANET OVERVIEW

In this section, the VANET system model, DSRC spectrum and the VANET characteristics are demonstrated in a comprehensive manner.

1.1.1 VANET System Model

The VANET system model is illustrated in Figure1.1, which consists of three major components namely, TA, fixed RSUs, and the on board units (OBUs) mounted on the moving vehicles. Each vehicle's OBU is connected with a group of sensors to gather the information such as velocity, breaking information and so on. These gathered information is sent as messages to surrounding vehicles via the wireless medium. All RSUs are interconnected with each other and are in turn connected to the TA through a wired connection. The TA has the duty of maintaining the entire VANET system.

Trusted authority: TA is responsible for the registration of RSUs, vehicle OBUs and the vehicle users. In addition to this, it is also responsible for verifying the authorisation of the real OBU ID of vehicles or ID of users in order to avoid malicious vehicles (Ghosh et al. 2009) entering into the VANET system. The TA is assumed with high computational power and sufficient storage capability. The TA also maintains the vehicle user's key values as secret. In addition to this, the TA can disclose the real ID of OBUs in the case of broadcasting malicious messages or malicious behaviouring.

Road side unit: RSUs are generally stationary devices that are fixed aside the roads or in dedicated places such as parking places or road intersections. Similar to an OBU, an RSU also has a transceiver, antenna, processor, and sensors. The RSUs are intentionally fixed along the roads in order to give services to vehicles. For example, an RSU may be fixed near a road intersection to control the traffic and to reduce accidents. Each RSU uses a DSRC radio based on IEEE 802.11p radio technology to access the wireless channel along with a directional antenna or an Omni-directional antenna. If an RSU wants to transmit a message to a specific location, a directional antenna is used. In addition to this, it may also equip with other network devices to make communications with the TA and other RSUs. The RSUs have storage capability for storing the information coming from the vehicle's OBU and the TA.

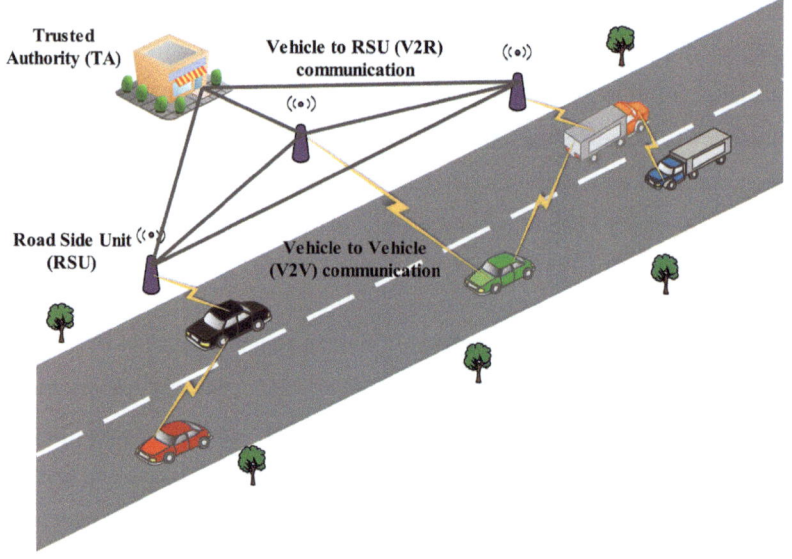

Figure 1.1 VANET system model

On board unit: OBU is a transceiver mounted on each vehicle to exchange information with RSUs and OBUs of other vehicles along with a computational device. The elements of an OBU are resource command processor for computation capability, read/write storage for storing and retrieving information, a user interface and a DSRC radio based on IEEE 802.11p radio technology to access the wireless channel. OBUs get power from the car battery. Furthermore, each vehicle has sensors like Global Positioning System (GPS) receiver, Tamper Proof Device (TPD), Event Data Recorder (EDR), speed sensor and forward and rear sensors to give input to the OBU. The sensors collect information about the local conditions of the vehicle. Among these devices, the GPS receiver is used to provide geographic information such as location information of the vehicle. The TPD is used to store sensitive data such as private key, group key and ID of the vehicle. These key values are used for secure communication in VANETs. The EDR is used to record information related to accidents or vehicle crashes. The speed sensor is used to collect the information such as velocity and breaking information. The forward and rear sensors are used to monitor the activities happening in front and back side of the vehicle. All these monitored and gathered information are sent as messages to neighbouring vehicles through the wireless medium.

1.1.2 Dedicated Short Range Communication (DSRC)

In October 1999, the Federal Communications Commission (FCC) allocated a spectrum of 75 MHz for DSRC (Zhu & Roy 2003). The 5.9 GHz DSRC spectrum has six service channels and a control channel with 10 MHz range. Orthogonal frequency division multiplexing (OFDM) modulation scheme is used in DSRC. The 75 MHz DSRC spectrum, as shown in Figure1.2 is divided into seven 10 MHz channels starting from channel number 172 (Ch 172) to channel number 184 (Ch 184). The Ch 178 is used entirely as the control channel that supports all power level safety application broadcasts. The Ch 172 and Ch 184 are dedicated channels for sending only safety messages and the other service channels (Ch 174, Ch 176, Ch 180 and Ch 182) are used for sending both non-safety and safety messages.

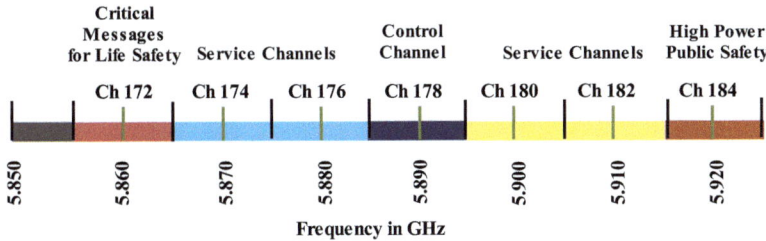

Figure 1.2 Channel layout for DSRC

1.1.3 VANET Characteristics

In VANETs, the wireless communication is basically executed in an ad-hoc manner. VANET has its own unique characteristics which are generally a mixture of wireless medium characteristics and ad-hoc characteristics. Some unique characteristics of VANETs are mentioned below:

Mobility: VANETs have relatively higher mobility than MANETs. Every node in VANET generally moves at high speed. Hence, the high mobility of nodes reduce the communication time between nodes in the network (Dhamgaye & Chavhan 2013 and Zeadally et al. 2012).

Dynamic network topology: VANET topology is changing quickly due to high mobility of vehicles and hence it is irregular in nature. This rapid change in network topology makes easy to attack the entire VANET network, and makes hard to notice the malicious vehicles.

Time critical: The information passed in VANET must reach the vehicle nodes within a particular time limit, so that decisions can be made by the vehicle node and the corresponding actions are taken promptly.

No power constraints: VANETs do not suffer from power issues as in mobile ad-hoc networks (MANETs), because the OBUs can get continuous power supply through the long life battery.

Variable network density: The network density of VANETs depend on the traffic density of vehicles, which can be low in rural and suburban areas and high during traffic jams.

Frequent disconnections: The vehicles communicate with the VANET network through the wireless medium. The dynamic topology, bad weather conditions and high density of vehicles may lead the vehicle nodes to frequent disconnections from the network.

Wireless communication: Since the wireless medium is used as the transmission medium in VANETs, the data transmission should be anonymous. If the wireless transmission medium is not protected securely, then anyone can malfunction with the same operating frequency (Blum & Eskandarian 2004).

Transmission power limitation: In the wireless access of vehicular environments (WAVE) architecture, maximum transmission power ranges from 0 to 28.8 dBm and the associated coverage distance may range from 10 m to 1 km. Therefore, the limited transmission power also limits the distance of the coverage area (Morgan 2010).

Wireless transmission impairments: DSRC wireless communication also has limitations on performance due to reflection, diffraction, refraction and scattering in urban surrounding areas.

Computing capacity and energy storage: VANETs do not receive energy or storage failure problems. However, large scaling environments require processing of large amount of information and it is absolutely a challenging issue.

1.2 SECURITY ISSUES IN VANET

There are many existing works (Kitani et al. 2008, Park & Lee 2012, Raya & Hubaux 2005, Lu et al. 2012, Lin et al. 2007, Lin et al. 2008, Hao et al. 2008 and Zhang et al. 2008c) that provide security facility in VANET. Among the existing works, some of the works (Lin et al. 2007, Lin et al. 2008, Hao et al. 2008 and Zhang et al. 2008c) focus on authentication

7

mechanisms. However, most of the existing authentication schemes are suffered from high computational cost and also that in some of the schemes the privacy is not provided for vehicle users during the authentication process.

To overcome the security issues in the existing authentication schemes, the authentication schemes proposed in this research work are computationally less expensive. The main objective of developing a dual authentication scheme is to improve the security in the vehicle side. The proposed dual authentication scheme depends on the vehicle secret key (VSK) which is given to the user during the time of registration by the TA and the fingerprint of the individual user. Even if the VSK value of any user is lost, the intruder cannot use that VSK for getting service from the TA. To prevent the intruder to use other users VSK, fingerprint of each authenticated user is included in the smart card issued by the TA. Moreover, an anonymous authentication scheme is proposed for authenticating the vehicle users and RSUs in an anonymous manner to preserve their privacy from malicious entities in the network.

Many existing schemes available in the literature for group key management in the wired and wireless networks (Cheng et al. 2015, Wong et al. 2000, Zheng et al. 2007, Zhou & Yong 2009, Naranjo et al. 2012, Vijayakumar et al. 2013). The main limitations of these existing works are that these schemes are more expensive in terms of computation and communication complexities which leads to the decrease in performance in that schemes. Moreover, the memory requirements are high in most of the existing schemes. To overcome these security problems, a dual key management scheme is proposed in this research work which is superior to the existing works in many ways. First, the computation complexity of the TA and VANET user is reduced substantially by minimizing the number of arithmetic operations taken by the TA and VANET user. In order to minimize the computation time in both the TA and vehicle side, the CRT based key management scheme is used. In addition, the number of computations is reduced by validating the credentials using intelligent agents in the OBU. Hence, the overall computing power is enhanced in each vehicle. Second, comparing with all the existing group key management algorithms, the number of key values stored by VANET users is also minimized in this work. Finally, the proposed algorithm reduces the amount of information needed to be communicated for updating the group key values when there is a change in the group membership.

1.3 PROPOSED WORKS

The primary step to ensure security in VANET is performed by providing an authentication mechanism through which it is easy to ascertain all the authenticated vehicles. Authentication is the process of verifying a user identity prior to granting access to the network. It can be considered as the first line of protection against intruders. The authentication process ensures that only valid vehicles can be part of the group in VANET. In this research work, a new dual authentication scheme is proposed to provide the security improvement in the vehicle's side to resist malicious users entering into the VANET. After completing the authentication process, the TA can multicast the information to the authenticated vehicles. The authenticated vehicles can broadcast that information to other vehicles in a secure way. To multicast the information from the TA side and to broadcast the information from one vehicle to other vehicles, a dual key management technique using Chinese Remainder Theorem (CRT) is developed in this research work. Moreover, to preserve the privacy of the users an anonymous authentication scheme is proposed in this research based on bilinear pairings. In addition, to prevent malicious users and malfunctioning OBUs continuing with the system, an efficient conditional tracking mechanism is also required for VANETs.

Similarly, each authenticated vehicle is required to authenticate the RSUs in an anonymous manner to avoid communication with malicious RSUs. Without a valid anonymous certificate and signatures, messages will not be accepted by the receivers. In order to avoid the man-in-the-middle attack during the key distribution a computationally efficient group key distribution (CEKD) scheme for secure group communication is proposed in this research work.

1.4 OBJECTIVES OF THE RESEARCH WORK

The major contributions of this research work are summarized as follows.

1. To propose a secure dual authentication technique with the capability of preventing malicious vehicles entering into the VANET system.

2. To develop a dual key management technique into the VANET to disseminate the information from the TA side to the group of vehicle users in an intelligent and secure way.

3. To develop a computationally efficient anonymous authentication scheme to verify the authenticity of OBUs without revealing their real identities.

4. To propose an efficient anonymous authentication scheme to check the legitimacy of the RSUs before getting the location based safety information (LBSI) from them.

5. To propose a secure key distribution mechanism to avoid man-in-the middle attack during the key exchange.

1.5 ASSUMPTIONS

Some important assumptions are considered in the proposed research works which are very essential for secure VANET communications. The assumptions are as follows.

- TA is powerful than vehicle OBUs and RSUs in terms of computation, communication, and storage capability.

- TA's public key is given to all vehicles and RSUs at the time of registration from which the vehicles and RSUs cannot find the TA's private key.

- TA has powerful firewalls and other protections that prevent them from being compromised.

- Each vehicle user keeps its credentials as secret which is given by the TA during the time of their registration. Similarly, each RSU keeps its credentials as a secret which is given by the TA during the time of its registration.

- The maximum group size of VANET group communication is considered to be 1000.

1.6 ORGANIZATION OF THE THESIS

Chapter 1 presents a brief overview of vehicular networks and the objectives of the proposed research work in detail.

Chapter 2 provides a detailed survey about the various related works about the security services and the security issues of VANET.

Chapter 3 depicts the architecture of the proposed research work.

Chapter 4 explains the dual authentication and dual key management schemes proposed in this research work for supporting secure group communication.

Chapter 5 details the computationally efficient privacy preserving anonymous authentication scheme proposed in this research work.

Chapter 6 describes the efficient anonymous authentication of an RSU proposed in this research work to efficiently authenticate the RSUs.

Chapter 7 provides the detailed research work about the computationally efficient key distribution scheme for securely distributing the key for group communication.

Chapter 8 summarizes the conclusions on this research work and suggests some possible future works related to this research work.

CHAPTER 2

LITERATURE SURVEY

2.1 INTRODUCTION

VANETs are the most hopeful approach to provide safety information and other infotainment applications to both drivers and passengers. VANETs are formed by intelligent vehicles equipped with OBUs and wireless communication devices. Hence, VANETs become a key component of the intelligent transport system. Even though VANETs are used in enormous number of applications, there are many security challenges and issues that need to be overcome to make VANETs usable in practice. A great deal of study has been done towards it, but security mechanisms in VANETs are not effective. This chapter provides a detailed survey on some major security attacks against various security services such as availability, confidentiality, authentication, integrity and non-repudiation and the corresponding countermeasures to make VANET communications more secure. Moreover, this chapter reviews the previous studies on various key management schemes for secure group communication in vehicular ad-hoc networks.

2.2 SECURITY SERVICES OF VANETS

Security is an essential requirement to be implemented in VANETs. The security services such as availability, data confidentiality, integrity, authentication, privacy and non-repudiation are used to measure the security of VANETs.

Availability: Availability ensures that the network is in a functioning state, and the users can get information at any time.

Confidentiality: Confidentiality guarantees that the data can be accessed only by designated recipient and other users are not permitted to access the data and thus it ensures that the data was not accessed until reception by the designated recipient.

Data integrity: Data integrity ensures that the content of the data is preserved from modification while in transit.

Authentication: Authentication is a mechanism to protect the VANETs against attacks due to the malicious entities in the network. Hence, it is considered to be the first line of defence against any kind of attacks in VANETs.

Non-repudiation: Non-repudiation is the service which ensures that the sending and the receiving parties of the data cannot deny its transmission and reception in the case of dispute (Hawi et al. 2009).

2.3   AVAILABILITY IN VANETS

Availability of the information is very essential because lack of availability feature may result in the reduction of VANET efficiency (Kassim et al. 2011). This section discusses about various threats on availability in VANETs. In addition to this, it also discusses about various existing works related to availability service.

2.3.1 Threats and Attacks on Availability

Several threats are available related to availability feature to affect the performance of the VANET system. The most commonly encountered threats on availability are summarised as follows:

- **Denial of service (DoS) attacks:** The DoS attacks can be performed by internal or external vehicles to the VANET (Zeadally et al. 2012, Malla & Sahu 2013 and He & Zhu 2012). In this attack, the attacker aims to chunk the main communication between vehicles and hence interrupt services. Therefore, the services cannot be accessed by the authorised users (Dhamgaye & Chavhan 2013).

- **Jamming attack:** In this attack, the attacker disrupt the radio wave base communications channel in the VANET system through the use of an over powered signal in the equivalent frequency range (Mpitziopoulos et al. 2009). This brings down the communication signal qualities until the frequency band becomes unusable or disconnected to the users.

- **Malware attack:** The malware (malicious software) attack can be penetrated into the VANET system via software components installed to operate the OBUs and RSUs (Dhamgaye & Chavhan 2013 and Al-kahtani 2012). Malware may lead to the disruption of ordinary functionality of the VANET system.

- **Broadcast tampering attack:** In this attack, the legitimate users may act as internal attackers and send false security alert messages in the VANET. This may conceal the correct safety messages to authorised users, which can lead to severe accidents. Moreover, it also affects the overall performance of the VANET system (Zeadally et al. 2012).

- **Blackhole attack:** The Blackhole attack usually takes place by a legitimate VANET user being compromised by external sources for a number of reasons. When a vehicle forwards the packet through this node, it silently drops the packet instead of relaying them to the recipients (Zeadally et al. 2012, Raj & swadas 2009, Misra et al. 2011 and Al-kahtani 2012).

- **Greedy behaviour attack:** In the greedy behaviour attack, the attacker attacks the functionality of the Message Authentication Code (MAC) layer of IEEE 802.11. The malicious vehicle intentionally abuses the MAC protocol to increase the bandwidth at the cost of other vehicles. The most important reason of this attack is to forbid other vehicles from the utilisation of support and services of the VANET. In the greedy behaviour, the malicious vehicle also tries to shorten its waiting time for quick access. This would cause a collision and burden problems in the wireless medium, which contributes to delay in the legitimate user's services (Raya & Hubaux 2004).

- **Spamming attack:** In this type of attack, the attacker injects unwanted spam messages such as advertisements in the VANET system and uses the bandwidth unnecessarily to cause voluntary collisions (Dhamgaye & Chavhan 2013 and Zeadally et al. 2012).

2.3.2 Works on Availability

A considerable number of protocols have been developed to improve the performance of the availability service. This section represents some of the most efficient existing approaches for the enhancement of the availability services in VANETs.

Kitani et al. (2008) proposed a new technique called 'message ferrying' to enhance the message propagation in low density areas. In this technique, buses are considered as message ferries that travel along usual routes. Buses can be used to collect as much traffic information as possible from vehicles in their surrounding location and then provide the collected information to neighbouring vehicles at regular intervals in order to enhance message propagation effectiveness in low density areas.

Okamoto & Ishihara (2010) proposed a zone-based forwarding (ZBF) messages method for information propagation in VANET. In this method, the entire network area is divided into predefined effect-areas, each of which is referred to as a zone. One vehicle in each zone is considered as a message forwarder and it is assigned to do the task of transmitting the warning messages at regular intervals to alert other vehicles in that zone. The main drawback of this method is the forwarder vehicle suffers with communication and computation overheads.

Akila & Iswarya (2011) proposed an efficient data replication method for data access applications in VANETs. Due to the high mobility of vehicles and the dynamic change of topology in VANETs, frequent disconnections may happen. If frequent disconnections occur, the vehicles are not able to access data from each other. The effect of the irregular connectivity problem has been reduced by the data replication method and enhances data access performance in distributed systems such as VANET.

Since, vehicles have limited storage space, they cannot replicate large data such as large music files or video clips. To overcome this problem, the vehicles are grouped together and form a platoon and the vehicles in a platoon are requested to contribute a portion of their buffers to replicate data for other vehicles in the same platoon and also share data with them. When a vehicle wants to leave from the platoon, it transfers its buffered data in advance to other vehicles in the platoon so that the other vehicles in the platoon can even access the data after it leaves. The limitation of this method is that frequent leaving and entering of vehicles in the platoon may incur additional computation overhead.

Park & Lee (2012) proposed an efficient method by using the data replica of RSU to offer rapid information delivery and therefore improve the data accessibility in VANETs. In this method, the data access pattern and driving pattern are used to select the data item which should be replicated in the RSU. The replicated data in the RSUs are requested to send directly to the vehicle without having communication with RSUs in which the original data is available. The main limitation of this method is the data replication process may take too much time if the size of the data is large.

2.4　CONFIDENTIALITY IN VANETS

Confidentiality is the protection of information so that unauthorised users cannot access it. The confidentiality of the information is very essential because a lack of confidentiality may result in the disclosure of sensitive data (Zaharuddin et al. 2010). This section represents various threats on confidentiality in VANETs. In addition to this, this section also explains about various existing works related to confidentiality.

2.4.1 Threats and Attacks on Confidentiality

There are several ways through which the confidentiality of the VANET system can be compromised. The most commonly encountered threats on confidentiality are summarised as follows:

- **Eavesdropping attack:** Eavesdropping is the intentional attempt against confidentiality to get information about protected data. Through this attack, the protected information is disclosed to unauthorised users in the VANET system which leads to information misuse such as identity theft and collection of location data of a target vehicle that can be used for tracking vehicles (Dhamgaye & Chavhan 2013).

- **Traffic analysis attack:** In VANETs, the traffic analysis attack is not an active attack. It is a passive attack against confidentiality of the users. In this attack, the attacker may listens the message transmission and then observes the frequency and duration of messages being sent. From the observations, the attacker can guess the nature of communication and tries to gather the maximum of valuable information for its personal purposes.

- **Man in the middle attack:** As the name suggests, the attacker can set up this attack in the middle of V2V or V2R communications to monitor and change the messages. In this attack, the attacker has a control over all the V2V or V2R communications, but communicating parties think that they are directly communicating to each other by the way of a private connection (Das et al. . 2006).

- **Social attack:** In this attack, the attacker sends unethical and unmoral messages to vehicle drivers so that the drivers get confused and upset. The primary aim of the attacker is to get the rightful vehicle users react in an annoyed manner after having such kind of messages and hence affect the driving performance of the vehicle in the VANET system (Raya & Hubaux 2005).

2.4.2 Works on Confidentiality

There are few existing works available for confidentiality to protect data that contains sensitive information from unauthorised users. This section presents the existing approaches for providing confidentiality in VANETs.

Sun et al. (2010a) proposed an ID-based security system in which confidentiality of sensitive information is protected by using symmetric or public key encryptions, for secure data communications. It is to be noted that, safety-related messages need not be encrypted because they do not contain sensitive information. However, confidentiality oriented messages are important where vehicles obtain data from internet services and from RSUs.

Lu et al. (2012) proposed an efficient key management scheme to achieve data confidentiality in location-based services (LBSs) in VANETs. The confidentiality should be carefully maintained to protect service contents from passive eavesdroppers in VANETs. In this approach, if a vehicle user does not join in the VANET system currently, then he/she cannot access the current LBS contents. In order to achieve confidentiality in an LBS session, all joined vehicle users share a secure session key which is utilised to encrypt service contents. The major limitations of all these schemes is that the computation complexity is high.

2.5 AUTHENTICATION IN VANETS

Authentication is a mechanism to protect the VANETs against attacks due to the malicious entities entering into the system. It can be considered as the first line of protection against attackers. The authentication process in VANETs protects the rightful nodes from the interior and exterior attackers (Ameneh & Ghaffar 2013). This section discusses the various threats on authentication in VANETs. Besides this, it also gives details about various existing works related to authentication.

2.5.1 Threats and Attacks on Authentication

Several threats are present related to authentication to affect the performance of the VANET system. The most commonly encountered threats on authentication are summarised as follows:

- **Sybil attack:** In the Sybil attack, the malicious entity can act as multiple IDs at once. In this attack, an attacker broadcast numerous messages with different IDs to other vehicles. The receiving vehicles think that these messages are broadcasted from different vehicles and hence they feel that there is traffic jam and they are enforced to change their routes to make the road clear. The Sybil attack is very difficult to identify and it is really dangerous to the VANET environment (Douceur 2002, Grover et al. 2011 and Park et al. 2009),

- **Tunnelling attack:** The tunnelling attack is most alike to the wormhole attack (Zeadally et al. 2012). In this attack, the external wormhole attackers connect two far-away parts of the vehicular network by using an extra communication channel called as a tunnel (Rawat et al. 2012). Therefore, the users from two far-away parts of the network can communicate as neighbours.

- **GPS spoofing:** A location table is maintained in the GPS satellite, which contains the geographic location information about all the vehicles on the VANET. An attacker can produce false readings about its position rather than telling its correct position using the GPS satellite simulator to fool the vehicles to think that it is available in a different location (Wen et al. 2011).

- **Impersonation attack:** An impersonation attack is an attack in which an attacker successfully guesses the ID of one of the rightful users in the VANET system Al-kahtani (2012).

- **Replay attack:** Replay attack is also known as a playback attack in which a valid data is maliciously or fraudulently retransmitted or delayed to produce an unauthorised effect. To avoid replay attack, the VANET requires a time source with a cache memory to compare the recently received messages with the newly received messages.

- **Free-riding attack (or active free-riding attack):** This kind of attack is created by an active malicious user who is involved in the cooperative authentication communication by making false authentication efforts. In this attack, the attacker verifies the authentication efforts of other vehicle users and unites them to create an authentication effort. By doing this, the attacker provides a valid verification effort more intelligently which is verified by others. This kind of attack is very hard to sense by nearby users and the TA (Lin & Li 2013).

- **Masquerading:** In this attack, the attacker actively pretends to be another vehicle by using false IDs. This attack takes place when one user makes belief to be a different user to gain unauthorised access through legitimate access identification.

- **Key and/or certificate replication:** In this attack, the attacker uses duplicate keys and certificates of other vehicles as an authentication proof to create doubt which makes hard to TAs for identifying a vehicle. The main objective of this kind of attack is to confuse TAs and avert authentication of vehicle users in hit-and-run events.

- **Message tampering:** In this attack, the attacker is altering the messages exchanged in V2V or V2R communication so as to forge transaction application requests or to counterfeit responses.

2.5.2 Requirements for Authentication

The most important prerequisites for offering dependable and efficient authentication in VANETs are listed below:

- **Computational overhead:** The number of cryptographic operations to be performed by a vehicle node or RSU or TA for verifying an authentication request should be minimised. For instance, the time required for checking a digital signature should be minimised.

- **Bandwidth utilisation:** The bandwidth should be effectively utilised by vehicles (in bytes per second) for an authentication request in the case of exchanging cryptographic secret keys and credentials.

- **Response time:** The time required to reply for an authentication request should be minimised.

- **Strong authentication:** Authentication techniques must be stiff enough to prevent attackers.

- **Scalability:** Authentication process should be in a scalable manner.

2.5.3   Works on Authentication with Privacy Preservation

Authentication is the first line of defence against any kind of attacks which verifies the legitimacy of a vehicle user prior to granting access to the network. The message authentication is the way through which the vehicle users can differentiate bogus information from reliable information and hence oppose modification attacks and impersonation attacks. Integrity verification and ID verification are two important security checks to be performed in message authentication. Based on the verification pattern of messages, there are two classes of authentication schemes namely one-by-one message verification (Lin et al. 2007, Lin et al. 2008, Hao et al. 2008 and Zhang et al. 2008c) and batch verification (Zhang et al. 2008b) and Shim (2012) schemes. Privacy is a mechanism to protect sensitive information about the vehicles or passengers from the intruders or attackers. The information about vehicle's circumstances may possibly affect its driver privacy. In the field of VANETs, bulk of the research works has been concentrated on privacy to guarantee security. The majority of existing works mainly depend on pseudonym-based approaches to effectively protect the privacy of vehicles. The pseudonym-based approaches have a common advantage to provide efficient privacy preservation. However, to avoid privacy-related attacks, the TA is required to change the pseudonyms in a frequent manner. Therefore, several privacy schemes use bilinear pairing-based elliptical curve cryptosystem (ECC) as their fundamental building block to avoid privacy-related attacks. The privacy is classified into two types:

User privacy: Protecting user's personal information from the other users or intruders in the network.

User location privacy: Protecting user's location information such as the location of a vehicle at a given time, or the route followed by a vehicle during a period of time are considered as personal data.

Some of the most important existing works for providing authentication with privacy preservation in VANETs are summarised and explained in this subsection.

Lu et al. (2008) proposed an efficient conditional privacy preservation protocol (ECPP) scheme which is based on the use of bilinear maps to attain user privacy and location privacy for the vehicles. In this scheme, an RSU provides multiple anonymous keys for a vehicle to thwart its communication from being traced. The primary limitation of ECPP scheme is that the RSUs are suffered from high latency during the pseudonym generation process. In addition, the vehicles mainly depend on RSUs to get their pseudonyms and corresponding keys for communication.

Moreover, in this scheme, the pseudonyms generated by the RSUs are required to inform to the TA before issuing to the vehicles. Since RSUs are usually vulnerable to physical attacks and easily compromised, the pseudonyms generation at RSUs is not effective.

Lin et al. (2011) proposed a Social-Tier-Assisted Packet (STAP) approach to attain receiver's location privacy preservation in VANETs. In this approach, the places such as well-traversed shopping malls and busy intersections of a city are considered as social sports and through which the social tier is formed. In STAP, the packet transmission is only done in the social sports. In this approach, each vehicle broadcast the packets in the social tier in order to enhance the packet delivery performance and also preserve location privacy of the receiving vehicles. Later, once the receiver enters into the region of one of the social spots, it can effectively pick up the package.

Lin et al. (2007) proposed a group signature and ID-based Signature (GSIS) scheme which is a secure and privacy-preserving protocol for VANETs which is based on the role of public key credentials. In this approach, the digital signature technique is used for V2V communications because everyone receives the right to experience the substance of the message and then retains the confidentiality of each message in V2V communications is not necessary. Thus, any recipient can authenticate the received message and make sure of the integrity and genuineness of the messages with the non-repudiation property. In the case of V2R communication, a signature scheme using ID-based cryptography is taken in the RSUs to digitally sign every message launched by the RSUs to certify its authenticity in order to reduce the signature overhead. Shen et al. (2013) developed a Cooperative Message Authentication Protocol (CMAP) for a two-dimensional (2D) city road scenario. In this protocol, each safety message holds the location data of the sender vehicle (which is made by utilising the GPS device). Message verifiers for each safety message are defined based on the locations of the sender. The selected verifiers check the legality of the message and at the same time other vehicles depend on the verification results from these verifiers. Particularly, the vehicles computation load is alleviated in CMAP because the verifiers share their verification results to other vehicles in a cooperative way, thus the number of safety messages that each vehicle requires to validate reduces considerably. In this method, the RSU validated messages one by one and transmitted the authentication result as a 128 bit hash value for each valid message, which increases the communication overhead heavily. Verifier's selection is based on the location of the sender at that time non-verifiers waited for the results of verifiers. Referable to the precariousness of the conveyance speed and route conditions, the CMAP are faced querying.

Zhang et al. (2008a) proposed a novel RSU-aided message authentication scheme named RAISE. In this scheme, a secret key is shared between the sender and RSU to generate a MAC. Each vehicle transmits the messages by signing it using the MAC. By receiving the messages, the vehicles cannot directly verify the messages signed with the MAC since the key to the MAC is not known to the receiving vehicle. However, the RSUs know the key to the MAC and hence it can verify the authenticity of the messages and notify the authentication results back to the neighbours' vehicles. The authors show that the performance of RAISE is better to that of conventional PKI (public key infrastructure)-based system, but the chief limitation of this approach is that it is heavily reliant on RSUs.

Shim (2012) proposed a conditional privacy - preserving authentication scheme, called CPAS for secure V2R communications in the VANET. In the CPAS, the messages are signed with pseudonyms to protect privacy of vehicles and fortify the fastest batch verification process at the RSUs. In this scheme, the RSUs have the capability of verifying multiple received signatures simultaneously. Therefore, the total verification time is greatly reduced. However, the TA can only recover the actual ID of a vehicle from any pseudo-ID. The major drawback of this scheme is that it takes large computational overhead. Since, the signature verification is based on bilinear pairing operation and the use of private key generator (PKG) is necessary to produce the VANET user's private keys.

Zhang et al. (2008b) introduced an ID-based batch verification (IBV) scheme for V2R communications in vehicular sensor networks (VSNs). This approach is the most efficient approach for VSNs to avoid any potential malicious attacks and resource abuse in a digital signature strategy. In this approach, the RSUs can authenticate several received signatures simultaneously and therefore the overall verification time can be reduced considerably. This approach utilised ID-based cryptography for generating secret keys for pseudo-IDs and thus it avoids the use of certificates. Hence, the transmission overhead can also be reduced greatly in this approach. However, in a large scale environment, the RSU would be hard to verify the large quantity of signatures within 300 ms and thus the scalability problem appears immediately.

Huang et al. (2011b) proposed a pseudonymous authentication based conditional privacy (PACP) scheme based on the use of pseudonyms. The pseudonyms are fake IDs which are not known to other entities in the VANET. The pseudonyms are used to communicate with other vehicles in the network in order to conceal the real ID of the vehicles. Since it hides the real ID, the anonymity of vehicle users is maintained. In this scheme, an efficient revocation mechanism is introduced in the TA to identify and revoke vehicles from the network in case of any dispute. If the vehicle is found as malicious, then its privacy is revoked and its anonymity is disclosed to other vehicles. Therefore, this method provides conditional privacy preservation to the vehicles in the system. The main drawback of this scheme is that the use of bilinear pairing operation takes large computational overhead.

Wasef & Shen (2009) proposed a message authentication acceleration (MAAC) protocol in which a CRL (certificate revocation list) is used which contains a list of certificates of all revoked vehicles and it is maintained by the TA. In the PKI, the TA first verifies the revocation status of a sender and then it verifies the sender's certificate and the sender's signature to authenticate the vehicle. In the initial phase of authentication, the TA checks the CRL to know the revocation status of the sender. This would cause long delay based on the size of the CRL. To overcome this limitation, the MAAC protocol replaces the CRL verification process by an efficient revocation check process by utilising an expeditious and secure Hash message authentication code (HMAC) function. Hence, the MAAC protocol reduces the authentication delay due to the CRL verification process in VANETs.

Jia et al. (2013) proposed an efficient privacy-preserving authentication scheme (EPAS). In this scheme, an efficient ID-based signature scheme is used to satisfy conditional privacy obligations via software solution. The EPAS scheme is suggested based on the utilisation of the disaster relief authority (DRA) instead of RSUs. The vehicles can register themselves with the DRA via emergency communication vehicles such as ambulances or fire trucks that travel into the disaster-affected region. In order to provide the conditional privacy, a special secret key is shared between the vehicle and the DRA, so that the DRA can track real ID of vehicles from the pseudonyms at the time of misbehaviour.

In addition, the lightweight signature and batch verification process are combined to reduce the computation and communication cost in order to offer fast and efficient authentication cost. This system also provides a vehicle group communication approach to allow vehicles to authenticate each other in the same group in an effective fashion. Even though there are many existing schemes (Lu et al. 2009, Tan 2010, Sun et al. 2010b and Lu et al. 2011) available to provide the authentication facility in VANET, most of the schemes suffer from huge message loss. In addition, anonymous authentication is not provided in some of the schemes (Mershad & Artail 2013, Wasef & Shen 2009, Zhu et al. 2008, Zhang 2015, Wasef et al. 2008 and Raya et al. 2006). The comparison of various authentication techniques is given in Table 2.1.

Table 2.1 Comparison of authentication techniques

Schemes	Communication Model	Cryptographic Method	Conditional privacy preservation	Batch Verification	Group Communication
GSIS	V2V &V2R	Group Signature and Identity (ID)-based Signature	Yes	No	No
CMAP	V2V	Group Signature	Yes	No	No
RAISE	V2V (RSU Based)	Symmetric MAC code based message signature.	Yes	No	No
CPAS	V2R	pseudo-IBS scheme with PKG	Yes	Yes	No
IBV	V2R	PKI-based signature scheme & pairing-based Cryptographic scheme	No	Yes	No
STAP	V2R	Pseudo-ids based signature scheme	Yes	No	No
PACP	V2V &V2R	Identity-based encryption (IBE) scheme & BLS short signature scheme	Yes	No	No
MAAC	V2V &V2R	Hash Message Authentication Code (HMAC)	No	No	No
EPAS	V2V	Identity based Cryptography	Yes	Yes	Yes

2.5.4 Computational Cost for Various Authentication Schemes

In this section, the performance of various privacy preserving authentication schemes is evaluated in terms of computational cost for certificate and signature verification process. The computational cost for certificate and signature verification process is the total time required to verify either one signature and one certificate or n signatures and n certificates to authenticate a vehicle and to check the integrity of a message. BLS (Boneh et al. 2003), ECPP (Lu et al. 2008), CAS (Gong et al. 2007), GSB (Lin et al. 2007) and KPSD (Lu et al. 2012) are proposed for privacy preserving authentication in VANETs based on the use of bilinear pairings. In these schemes, each vehicle reloads a large number of anonymous short time keys. Each key is used for a short period of time and then it will be discarded. The TA maintains the key distribution records to trace the possible malicious vehicles. The limitation of these schemes is that, each vehicle should store a large number of anonymous keys, where the user privacy and vehicles location privacy is difficult to break. However, these schemes consume more time during anonymous certificate verification process. Let T_p is the time required for performing a pairing operation, T_h is the time required for performing a hash function and the time required for performing the one point multiplication is T_m. For the aforementioned operations, the Type-A curve defined in the PBC (PBC 2005) library, is used with the default parameters (Lin et al. 2007). According to Lu et al. (2008), the time parameters T_p, T_h and T_m is found to be equal to 4.5 ms (milliseconds), 2.7 ms, and 0.001 ms, respectively. The time needed to perform exponentiation in groups G1 and G2 are denoted as T_{ep-1} and T_{ep-2} and are found to be equal to 0.6 ms. In Table 2.2, we summarize the certificate and signature verification cost for BLS, ECPP, CAS, GSB and KPSD schemes.

Table 2.2 Certificate and signature verification cost of various schemes

Method	For one Certificate & Signature	For n Certificates & Signatures
BLS	$4T_p + 2T_h$	$(2n + 2)T_p + 2nT_h$
ECPP	$3T_p + 11T_m$	$3nT_p + (10 + n)T_m$
CAS	$5T_p + 2T_h$	$(4n + 1)T_p + 2nT_h$
GSB	$3T_p + 4T_{ep-1} + 5T_{ep-2}$	$3nT_p + 4nT_{ep-1} + 5nT_{ep-2}$
KPSD	$4T_p, +5T_{ep-1} + 5T_{ep-2}$	$(3+n)T_p, +(4+n)T_{ep-1} + 5nT_{ep-2}$

It can be seen that T_p and T_h are the largest time-consuming operations in the signature verification process. Since, BLS and GSB schemes require more time for their signature verification process when compared to CAS, KPSD and ECPP schemes. The total verification cost of BLS, ECPP, CAS, GSB and KPSD schemes for a single certificate and signature are

$$T_{verify}^{BLS} = 4T_p + 2T_h = 4 * 4.5 + 2 * 2.7 = 23.4 \ ms$$

$$T_{verify}^{ECPP} = 3T_p + 11T_m$$

$$= 3 * 4.5 + 11 * 0.001 = 13.511 \ ms$$

$$T_{verify}^{CAS} = 5T_p + 2T_h$$

$$= 5 * 4.5 + 2 * 2.7 = 27.9 \ ms$$

$$T_{verify}^{GSB} = 3T_p + 4T_{ep-1} + 5T_{ep-2}$$

$$= 3 * 4.5 + 4 * 0.6 + 5 * 0.6 = 18.9 \ ms$$

$$T_{verify}^{KPSD} = 4T_p + 5T_{ep-1} + 5T_{ep-2}$$

$$= 4*4.5 + 5*0.6 + 5*0.6 = 24\ ms$$

From the above calculations, it can be seen that, the verification time of ECPP scheme is lower than that of BLS, CAS, GSB and KPSD schemes.

2.6 DATA INTEGRITY IN VANETS

Data integrity is one of the security services in VANETs which concentrates on retaining and promising the exactness and reliability of information over its entire transmission. The failure in data integrity has also been a failure of data security. This section gives details about various threats on data integrity in VANETs and also it gives explanation about various existing works related to data integrity.

2.6.1 Threats and Attacks on Data Integrity

There are many ways through which the data integrity service can be compromised. The most commonly available threats to data integrity are given below:

- **Masquerading attack:** In this attack, an attacker enters into the VANET system as a valid user through stolen passwords and attempts to create false messages that cause the appearance of receiving from a valid user. For instance, a malicious vehicle may act as an emergency vehicle and therefore fool the other vehicles to slow down their speed (Gamage et al. 2006).

- **Replay attack:** In a replay attack (Parno & Perrig 2005), the attacker maliciously or fraudulently repeat or delays a valid data transmission and thus take the benefit of the valid user. For example, the sender vehicle A requires proving the ID to the receiver vehicle B. The receiver vehicle B requests the sender vehicle A to provide its password as a proof of ID. In the intervening time, the attacker spies the conversation and keeps the password of the sender vehicle A. After the exchange is over, the attacker connects to the receiver vehicle B and shows the ID proof of sender vehicle A through which the receiver vehicle B accepts the request and provides the authentication to the attacker.

- **Message modification attack:** In this attack, an attacker either modifies or alters some part of the existing data message to be transmitted and therefore it brings about an unauthorised effect (Rawat et al. 2012). For example, the attacker alters the data as the route is congested for clear roads, and thus making the vehicle users to change their routes.

- **Illusion attack:** In this attack, incoming data from antennas and data collected by invalid sensors about traffic warning messages based on the current road condition create an illusion to vehicles as its neighbourhood (Lo & Tsai 2007). Vehicle accidents and traffic jams maybe easily caused by the illusion attack and also degrades the performance of the VANET system in terms of unwanted bandwidth utilisation.

2.6.2 Works on Data Integrity

In order to protect the integrity of the transmitting message, digital signatures are generated and attached with the messages (Guo et al. 2007). Some of the existing works refers to retain the exactness and reliability of information for preserving integrity in VANETs is explained in this section.

Lin et al. (2011) proposed a STAP approach to get receiver's location privacy preservation in VANETs. In addition to this, they were also concentrated on providing data integrity to enhance the data security. In order to achieve data integrity, the sender first computes the MAC and attaches this code to the message and then sends it to the receiver. On receiving this message, the receiver uses their session keys to check the MAC in order to achieve integrity.

Lin & Li (2013) proposed an efficient cooperative message authentication scheme in VANETs. This scheme is concentrated mainly on authentication. Besides that, this scheme also concentrates on preserving the integrity of the transmitted message. In order to achieve the integrity of the transmitted message, the sending vehicle digitally signs the messages to be sent using its corresponding private key. Then, the other vehicles utilise the sender's public key associated with this signature to verify the integrity of the message.

Lin et al. (2007) proposed a GSIS scheme. This system uses the digital signature technique to sign each message transmitted by the vehicle users and the RSUs. Thus, any recipient can confirm the received messages and make sure of the integrity.

2.7 NON-REPUDIATION IN VANETS

Non-repudiation guarantees that the transmitter and recipient of a message cannot later deny the transmitted message and the received message. This part discusses about the threat related to non-repudiation in VANETs and also it presents various existing works related to non-repudiation. The non-repudiation service can be compromised by a repudiation attack.

2.7.1 Attack on Non-repudiation

- Repudiation attack: In this attack, an attacker may deny the participation of sending and receiving of messages and therefore create confusion for the audit entity such as the TA. Some of the existing works for providing non-repudiation service in VANETs are explained in this section.

2.7.2 Works on Non-repudiation

Li et al. (2014) proposed a new technique ACPN (a novel authentication framework with conditional privacy-preservation and non-repudiation) for VANETs. In ACPN, the non-repudiation is achieved through the public key cryptography (PKC) based pseudo-identities and digital signatures. If a malicious vehicle transmits a fraudulent message and denies the transmission of that message, then the TA has the capability to open the signature in the message to reveal the actual ID of the vehicle. The main advantage of this method is reduced computation cost. The main limitation of this method is that certificate management is difficult.

Choi & Jung (2009) proposed an ID-based cryptosystem to provide non-repudiation and also to avoid the overheads of certificate management in PKI-based cryptosystem. In this scheme, timestamps are used which contain the composed date and time of a message and hence confirms that the message will be valid only for a certain period of time. However, these systems cannot promise strong non-repudiation because these systems suffer from the inherent weakness like the key escrow problem of an ID-based cryptosystem. (Biswas & Misic 2013) also proposed a cross-layer approach to privacy-preserving authentication in WAVE-enabled VANETs. In this approach, they concentrated on providing non-repudiation to prevent the repudiation attack. In order to provide non-repudiation, a signature is created for each message. To sign a message, a sender requires a unique secret key and session parameters. Eventually, once the message is signed and delivered, then the sender of the message cannot deny the signature for the broadcast message.

2.8 COUNTERMEASURES ON VARIOUS SECURITY ATTACKS

In Table 2.3, all the existing security attacks in VANETs are summarized and for each attack, the violated security services along with the related possible countermeasures are described.

Table 2.3 Security problems and their countermeasures

Attack	Countermeasures	Violated Security Services
Denial of Service attacks	✓ use pre-authentication scheme (He & Zhu 2012) ✓ Use An Attacked Packet Detection Algorithm (APDA) (Roselinmary et al. 2013)	✓ Availability
Jamming attack	✓ Use threshold based jamming countermeasure (TJC) technique (Mpitziopolos et al. 2009) ✓ use the frequency hopping technique (Malla & Sahu 2013)	✓ Availability
Malware attack	✓ Installing anti-malware applications ✓ Installing basic internet security applications	✓ Availability
Broadcast tampering attack	✓ Use digital signatures	✓ Availability
Spamming attack	✓ Embedded anti-malware frameworks	✓ Availability
Greedy behaviour attack	✓ Use watchdogs (Marti et al. 2000) ✓ Use intrusion detection systems	✓ Availability
Blackhole attack	✓ Use Detection, Prevention & Reactive AODV (DPRAODV) protocol (Raj & Swadas 2009) ✓ Use BAMBi technique (Misra 2011)	✓ Availability
Eavesdropping attack	✓ Use encryption techniques	✓ Confidentiality

Table 2.3 (Continued)

Attack	Countermeasures	Violated Security Services
Traffic analysis	✓ Use encryption techniques ✓ Use VIPER technique for V2I communications	✓ Confidentiality
Man in the middle attack	✓ Use strong authentication methods ✓ Use hash functions	✓ Confidentiality
Social attack	✓ Use digital signatures	✓ Confidentiality
Sybil attack	✓ Use a timestamp series approach (Park et al. 2009) ✓ Use a distributed & robust approach (Grover et al. 2011)	✓ Authentication
Masqurading	✓ Don't save passwords on computers ✓ Change passwords on a regular basis ✓ Use long and hard to guess passwords	✓ Authentication & Data integrity
Replay Attack	✓ Use a globally synchronized time for all nodes ✓ Use time stamps or nonce	✓ Authentication & Data integrity
Impersonation attack	✓ Use SPECS (Secure and Privacy Enhancing Communications Schemes) (Chima et al. 2011) ✓ Use digital certificates	✓ Authentication
Global Positioning System (GPS) Spoofing attack and tunnelling attack	✓ analyse of signal strength distribution between transmitted and received signals (Xiao et al. 2006)	✓ Authentication

Table 2.3 (Continued)

Attack	Countermeasures	Violated Security Services
Message tampering	✓ Use hashing techniques. ✓ Use digital signatures. ✓ Use strong authorization.	✓ Authentication
Key and/or Certificate Replication	✓ Use strong authentication. ✓ Avoid sharing of keys	✓ Authentication
Free-riding attack with fake authentication efforts	✓ Use strong authentication.	✓ Authentication
Message modification attack	✓ Use digital signatures	✓ Data integrity
Illusion attack	✓ Use trusted hardware	✓ Data integrity
Repudiation attack	✓ Use digital signatures ✓ Use PKC-based pseudo identities (Li et al. 2011)	✓ Non-repudiation

2.9 WORKS ON KEY MANAGEMENT

In this section, some of the existing group key management methods used in the wired and wireless networks are discussed in detail.

Among these schemes, Wong et al. (2000) presented a novel solution to the scalability problem of group or multicast key management. They introduced the concept of key graphs for specifying secure groups. In addition, they presented three strategies for securely distributing rekeying messages after a join and leave operation in the secure group. In the rekeying strategies, join and leave protocols have been implemented in a prototype key server that they have built. The main limitation of this approach is the increased computational complexity. Zheng et al. (2007) proposed two centralized group key management protocols based on the CRT. The main advantage of their approach is that the number of broadcast messages to distribute the group key to user side is minimized. Moreover, the user side key computation is also minimized. However, the main limitation of their approach is that computation complexity of the key server is very high. Zhou & Yong (2009) proposed a CRT based static key structure for distributing the group key to the members of the group when group membership changes. The main contribution of this work is that it minimizes broadcast messages and also minimizes user side key computation. However, it also increases the workload of key server by allowing the key server to find a common group key by using CRT for 'n' number of congruential equations. Naranjo et al. (2012) presented a new algorithm for key management to provide security and privacy. Vijayakumar et al. (2013) proposed a Greatest Common Divisor (GCD) based key distribution protocol that focuses on two dimensions. The first dimension deals with the reduction of computational complexity and second dimension aims at reducing the amount of information stored in the Group Center and group members while performing the update operation in the key content. The main limitation of these existing works is that the computation complexity involved in rekeying operations leading to the decrease in performance. In addition, the memory requirements are high in most existing schemes.

2.10 LITERATURE SURVEY GAPS

Even though, there were many existing schemes available in the literature for providing authentication and data integrity, the existing schemes are more expensive in terms of computation overhead. Similarly, the main limitation of these existing works is that the computation and communication complexities involved in rekeying operations leads to the decrease in performance. In addition, the storage complexity for storing the key values is also high in most of the existing schemes. Moreover, the anonymous authentication techniques that were explained in the literature for preserving the privacy of the vehicle users and RSUs are computationally more expensive. Finally, most of the key management schemes suffer from the man-in-the-middle attack during secure key distribution from the TA side to the user side.

2.11 PROPOSED WORK

To overcome the literature survey gaps, in this research work, first, a dual authentication scheme is proposed to provide a high level of security in the vehicle side to effectively prevent the unauthorized vehicles entering into the VANET. Second, a dual group key management scheme is proposed to distribute a group key to a group of users and to update such group keys during the users' join and leave operations in both computational and communicational efficient manner. Third, an anonymous authentication scheme for a vehicle user is proposed to preserve the privacy of the vehicle users in a computationally efficient way. Fourth, an anonymous authentication scheme for an RSU is proposed to preserve the privacy of the RSUs in a computationally efficient way during the transmission of location based safety information (LBSI) messages. Fifth an anonymous distribution technique is proposed to preserve the privacy of the vehicle users during the exchange of group keys from TA side to the vehicle users and to avoid man-in- the-middle attack.

2.12 CONCLUSIONS

VANETs are becoming the most challenging and promising research field in intelligent transportation systems due to its safety related and non-safety related services for passengers comfort. However, its high mobility, frequent topology change and several security attacks make the implementation of VANETs a big challenging issue. VANET is vulnerable to different sorts of security attacks due to its broadcast infrastructure. This chapter reviews the VANET system model, characteristics of VANETs, various security problems in VANETs, various security services for VANETs and various works on key management. In addition, this chapter summarised all the security attacks and the related possible countermeasures for easy understanding and readability. This chapter provides a clear insight about the importance of VANET security with respect to the security services and key management. The security of VANETs takes a vital role for basic life debilitating circumstances. For providing a secure communication among the millions of vehicle nodes and to ensure the safe achievement of various goals through several applications, a security architecture with suitable security services is essential for VANETs.

CHAPTER 3

SYSTEM ARCHITECTURE

The system architecture of this research work is depicted in Figure 3.1. The architecture consists of three major components namely Trusted Authority (TA), Road Side Units (RSUs) and Vehicles. The TA consists of five modules namely user registration module, authentication module, key management module, conditional tracking module and communication module.

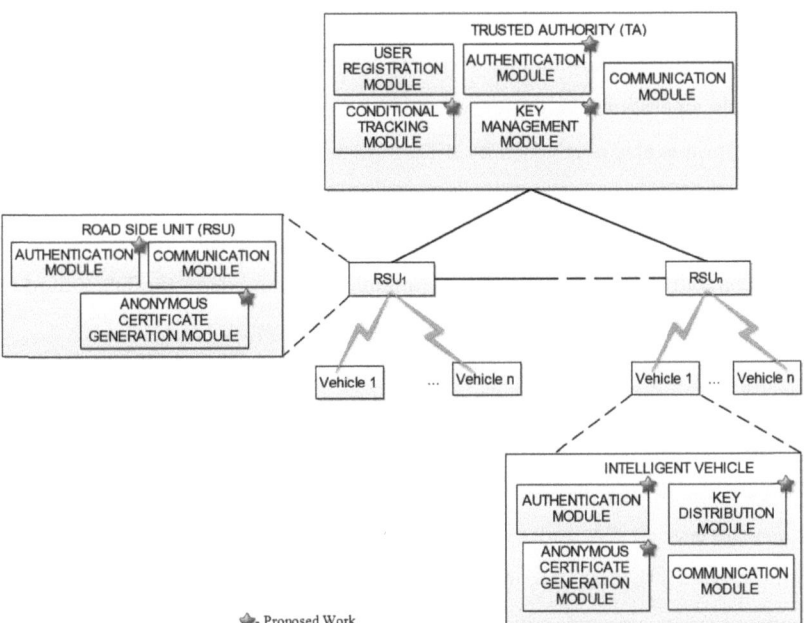

Figure 3.1 System architecture

The user registration module is responsible for user registration and initial key generation for each user after their successful registration in the VANET system. The user registration itself is divided into two types namely online registration and offline registration. In the online registration mode, the vehicles users are required to provide their personal details to the TA through TA's website. In the offline registration mode, the vehicle users are required to directly go the TA and provide their necessary documents to get the essential credentials to prove their legitimacy to other vehicles and RSUs. The authentication module is used to perform TA's authentication process during dual authentication process which is proposed in this research work. The key management module is responsible for key generation, key distribution and key updating process. The key management module is used to generate the group keys for two different levels of vehicle users such as primary users and secondary users and anonymous private and public keys to each user and RSUs for anonymous authentication process. The key updating process is carried out by the key management module during the time of user leave or user join operations in the group communication. Moreover, a secure key distribution is also performed by a TA to distribute the group key to the group of VANET users based on bilinear pairings. The conditional tracking module is used to efficiently track the malicious vehicles in the case of misbehavouring and dispute. The communication module in the TA is used to make communications with RSUs in the VANET system.

The RSU consists of three modules namely communication module, authentication module and anonymous certificate generation module. The communication module in the RSUs is used to make communication with the vehicles and TA and to transfer the data. The authentication module is used to verify the vehicles in the RSU's coverage region through anonymous authentication process which is proposed in this research work. In the anonymous authentication process, the vehicle users are authenticated by other users and RSUs without providing their real identities to them. By doing this kind of anonymous authentication, the privacy of vehicle users is preserved from other users in the VANET system. For performing the anonymous authentication, the vehicle users are required to generate the anonymous credentials using the parameters obtained from the TA during the time of registration. In order to generate the anonymous credentials an anonymous certificate generation module is used in each RSU. The anonymous certificate generation module is used to generate the anonymous certificates and signatures to prove RSUs legitimacy to vehicle users in an anonymous manner before sending the LBSI messages to them.

The vehicle user consists of four modules namely authentication module, key distribution module, anonymous certificate generation module and communication module. The authentication module in the intelligent vehicle is used to perform vehicle's authentication process during dual authentication process and to perform anonymous authentication which are proposed in this research work. In this proposed dual authentication scheme the vehicle secret key (VSK) and the fingerprint of the individual user are given to the user through a smart card during the time of registration by the TA. Even if the VSK value of any user is lost, the intruder cannot use that VSK for getting service from the TA. To prevent the intruder the fingerprint of each authenticated user is also included in the smart card issued by the TA. The anonymous certificate generation module is used to generate anonymous certificates to prove their legitimacy to other vehicles and RSUs in an anonymous manner. The key distribution module is responsible for sending the group key to the secondary users in the case of primary users.

CHAPTER 4

DUAL AUTHENTICATION AND DUAL KEY MANAGEMENT FOR GROUP COMMUNICATION

4.1 INTRODUCTION

VANET is a distributed, self-organizing communication network, which is built among moving vehicles. Due to the promising features and their security properties, VANETs have extensive attention in the research community in recent years. The TA provides a variety of online premium services to the VANET users through RSUs. Various statistical studies reveal that due to road accidents, many people have either died or injured and the traffic jams generate a tremendous waste of time and fuel. In order to solve these problems and to enhance the driving comfort, appropriate traffic information should be provided to the drivers in a smart and secured way. Therefore, VANETs are developed to provide attractive services such as safety services that include curve speed warnings, emergency vehicle warnings, lane changing assistance, pedestrian crossing warnings, traffic-sign violation warnings, road intersection warnings and road-condition warnings. In addition, it can offer the comfort services such as weather information, traffic information, location of petrol stations or restaurants, and interactive service such as Internet access. Even though, these services make driving comfort, the Intelligent Transport System (ITS) technology heavily depends on the intelligent security and privacy-preserving protocols to enhance the quality of experience for the drivers and passengers without fear for their safety and personal privacy (Wischhof et al. 2005 & Shen et al. 2014). Since, V2V and V2R communications are performed through an open wireless channel, these communications are vulnerable to various kinds of attacks such as interference, eavesdropping, jamming, etc. (Dhamgaye & Chavhan 2013).

The primary step to ensure security in VANET is performed by providing an authentication mechanism through which it is easy to ascertain all the authenticated vehicles (Busanelli et al. 2011, Huang et al. 2011a & Hao et al. 2011). Authentication is the process of verifying a user identity prior to granting access to the network. It can be considered as the first line of protection against intruders. The authentication process ensures that only valid vehicles can be part of the group in VANET. In this research work, a new dual authentication scheme is proposed to provide the authentication facility in the vehicle's side to resist malicious users entering into the VANET. After completing the authentication process, the TA can multicast the information to the authenticated vehicles. The authenticated vehicles can broadcast that information to other vehicles in a secure way. To multicast the information from the TA side and to broadcast the information from one vehicle to other vehicles, a dual key management technique using Chinese Remainder Theorem (CRT) is proposed in this research work. In this technique, the TA generates two different group keys for two different groups of users, namely primary user group and secondary user group. In the generated group keys, one group key is used for multicasting the information from the TA to primary users (PUs) and the other group key is issued for broadcasting the information from primary users to secondary users (SUs). However, the shared cryptographic group keys should be refreshed through a proper racing operation at the time of group membership changes due to new users joining into the network or old users leaving from the network. Therefore, an old group member has no access to present communications which provides forward secrecy. Similarly, a new member has no access to previous communications which provides backward secrecy. The proposed dual group key management scheme minimizes the computational cost of the TA and the group members in the rekeying operation. To achieve this goal, the TA performs only simple addition and subtraction operations to update the group key. Similarly, each vehicle user of the multicast group performs only one modulo division operation for recovering the updated key when the group membership changes.

4.2 PROPOSED DUAL AUTHENTICATION TECHNIQUE

This section explains our proposed dual authentication technique, which is used for secure VANET communication. To provide secure, authenticated communication in VANETs, initially, the TA selects two large prime numbers p and q. The value p helps in defining a multiplicative group z_p^* and q is used to fix a threshold value to select the group key values. Initially, the TA selects the $VSK_i (1 \leq i \leq n)$ from the multiplicative group z_p^* for 'n' number of vehicles which are given to the vehicle users when they complete the registration process. This VSK is used for authenticating the vehicles when they enter into the VANET to start communicating with other vehicles and RSUs. In order to improve the authentication process, a dual authentication technique is proposed in this research work where the authentication process is performed two times. For the first time, authentication is done on the vehicle side and the second time, authentication is done in the TA side and hence the intruder has no possibility to enter into the VANETs. In the TA, the authentication is performed by verifying the Hash Code (HC) generated by the vehicle using their VSK_i. The authentication was performed on the vehicle side by verifying the fingerprint given by the user at the time of registration. The main objective of introducing dual authentication technique is that anyone who finds the VSK of a vehicle cannot enter into VANET communication as they cannot produce the fingerprint of the corresponding vehicle user. The same is true if the attacker has only the fingerprint of a vehicle user and does not have VSK of that vehicle. Therefore, the dual authentication technique provides more security because these two factors are required in order to authenticate the vehicles.

In VANETs, the registration can be performed in two ways, namely online mode and offline mode. In the online mode, each VANET user performs registration process by submitting his/her details in the TA's website through internet connection. In contrast to the online mode, the user goes to the TA's office to complete registration in the offline mode. In this approach, the registration is performed in the offline mode. After completing the registration process, each VANET user must complete a dual authentication process to get Authentication Code (AC) in order to send messages in VANETs. After receiving the authentication code, the vehicles are permitted to receive services from the TA and also vehicles can exchange information from one vehicle to other vehicles. Figure 4.1 shows the dual authentication process performed by both the vehicle and the TA. The following steps explain the process of dual authentication in both the vehicle and the TA.

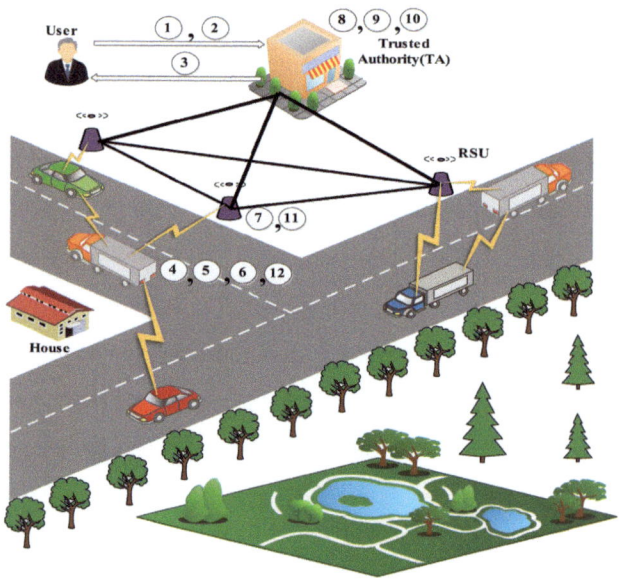

Figure 4.1 Authentication in vehicle and the TA

4.2.1 Registration Through Offline Mode

1) The VANET user first approaches the TA office directly to make offline registration and provide the essential information like name, address, phone number, email id etc. to the TA.

2) While each user performs registration, the TA gets the fingerprint of the corresponding user.

3) After completing the registration process, the TA provides the VSK_i to the registered user, which is unique for every vehicle and the TA also maintains the list of all the vehicles and their respective VSKs in its storage area. The TA provides the (VSK_i) to the user through a smart card which also contains the user's fingerprint which is given by the TA to the user after completing the registration process.

- The (VSK_i) is used for creating the Hash Code (HC) and the HC is verified by the TA for authentication and then the TA provides AC to the authenticated VANET users.

- The fingerprint is also verified in the vehicle side for authentication of the user during the time for making communication with the VANET.

- If the fingerprint is not matched with the fingerprint which is printed on the smart card, then the user is not permitted to make communications with VANETs.

4.2.2 Vehicle's Authentication Process

4) Each user store (VSK_i) in their vehicle in the tamper proof device which is equipped in the car. When a user wants to communicate with the VANETs, then the user first enters his/her fingerprint through a fingerprint device which is equipped inside the car. Then, the OBU of the vehicle compares this fingerprint with the smart card fingerprint that is already stored in the smart card.

- If they match, the user is allowed to communicate with TA and with the other vehicles.

- If they do not match, the user is not allowed to communicate with other users of the VANET.

4.2.3 Trusted Authority's Authentication Process and the Provision of Authentication Code (AC)

5) Each vehicle selects a random number N. After selecting the random number, it successively creates a Hash Code (HC) using N and VSK by SHA_256 (Matusiewicz et al. 2005) algorithm.

$$HC = SHA\_256\ (VSK\ \|\ N) \quad (4.1)$$

6) The vehicle encrypts the random number N, the Hash Code (HC) and the Vehicle ID, with its VSK, and broadcasts along with the Vehicle's identity ID_V, TA identity ID_{TA} and the time stamp TS_1 as shown in Equation (4.2). This forms the Authentication Request.

$$< E_{VSK}(N\ \|\ HC\ \|\ ID_V)\ \|\ ID_V\ \|\ ID_{TA}\ \|\ TS_1 > \quad (4.2)$$

Here, the identities of vehicle (ID_V), RSU (ID_{RSU}) and the TA (ID_{TA}) are dummy identities which are generated during the time of its registration using the following manner. In order to compute the dummy identities, the TA chooses two random numbers a_1 and b_1 such that $a_1, b_1 \in Z_q^*$ and computes $ID_V = g_1^{a_1} \times g_2^{b_1}\ mod\ q$. Here, g_1 and g_2 are the generators of Z_q^*. Similarly, the TA generates the dummy identities of RSUs (ID_{RSU}) and its dummy identity (ID_{TA}). The mapping from original identities to dummy identities is done only in the TA. The necessity of attaching the dummy identities in each message is to check the validity of the message source and the identification of the particular vehicle or RSU or TA. Even though these identities are disclosed to all vehicles, they do not reveal the privacy of the vehicle users or RSUs or TA. Because, even if these dummy identities are captured, they provide zero knowledge about the vehicle user or RSU.

7) The RSU receives the packet, appends its identity ID_{RSU} and increments the timestamp value TS_1 to get TS_2. Then, the RSU encrypts the entire message using the RSK which is known only to TA and RSU and forwards it to the TA as given in Equation (4.3).

$$< E_{RSK}(E_{VSK}(N \parallel HC \parallel ID_V) \parallel ID_V \parallel ID_{TA} \parallel TS_2 \parallel ID_{RSU}) > \quad (4.3)$$

8) The TA decrypts the packet received from RSU using RSK of the RSU and validates the RSU with its identity ID_{RSU} as given in Equation (4.4).

$$< D_{RSK}\big(E_{RSK}(E_{VSK}(N \parallel HC \parallel ID_V) \parallel ID_V \parallel ID_{TA} \parallel TS_2 \parallel ID_{RSU})\big) > \quad (4.4)$$

The TA also verifies its identity ID_{TA} after decrypting it using RSK. After verifying its identity, the TA decrypts the packet using VSK of the particular vehicle and verifies the ID_V

$$< D_{VSK}(E_{VSK}(N \parallel HC \parallel ID_V)) > \quad (4.5)$$

Then, the TA generates the HC using the random number N and the VSK by SHA_256 algorithm and then verifies the newly computed HC value with the HC which is sent from the vehicle side.

9) If the two HC values match, then the TA hashes the Hash Code to get the Authentication Code (AC).

$$AC = SHA\_256(HC) \qquad (4.6)$$

10) The TA includes the Vehicle ID, incremented time stamp value and also it includes the lifetime of the AC along with the AC and encrypts this sequence with its private key of TA ($TA-Pvt$) to create a digital signature. Therefore, any vehicle user can verify this digital signature using the public key of TA. But, no vehicle user can regenerate this digital signature because it is generated using the private key of the TA.

$$< E_{TA-Pvt}(AC \parallel ID_V \parallel TS_3 \parallel Lifetime)> \qquad (4.7)$$

This forms the authentication response. To securely transfer this AC to the appropriate vehicle user, the TA also encrypts this authentication response using the VSK value of the corresponding user and RSK of RSU.

$$< (E_{RSK}(E_{VSK}(E_{TA-Pvt}(AC \parallel ID_V \parallel TS_3 \parallel Lifetime))) \parallel ID_{TA}) > \qquad (4.8)$$

Finally, the TA sends the packet to the RSU.

11) RSU receives the packet from the TA and decrypts the packet using its RSK.

$$< D_{RSK}(E_{RSK}(E_{VSK}(E_{TA-Pvt}(AC \parallel ID_V \parallel TS_3 \parallel Lifetime))) \parallel ID_{TA}) > \qquad (4.9)$$

On receiving this message, the RSU is able to check the identity of TA(ID_{TA}), verifies that whether it is sent by the legitimate TA or malicious node. After verifying the identity of the TA, the RSU sends the packet to the vehicle user.

$$< (E_{VSK}(E_{TA-Pvt}(AC \parallel ID_V \parallel TS_3 \parallel Lifetime)) \parallel ID_{TA}) > \quad (4.10)$$

12) The vehicle decrypts the packet using its VSK, and then verifies ID_{TA}.

$$< D_{VSK}(E_{VSK}(E_{TA-Pvt}(AC \parallel ID_V \parallel TS_3 \parallel Lifetime)) \parallel ID_{TA}) > \quad (4.11)$$

After that, the vehicle verifies the ID_V by decrypting the resultant message using the public key of the TA.

$$< D_{TA-pub}(E_{TA-Pvt}(AC \parallel ID_V \parallel TS_3 \parallel Lifetime)) > \quad (4.12)$$

13) The vehicles then start sending the safety messages to other vehicles with this AC by encrypting the payload (message) using vehicle's group key k_{pug} or k_{sug} as shown in Equation (4.13).

$$< E_{k_{pug}}(payload) \parallel (E_{TA-Pvt}(AC \parallel ID_V \parallel TS_3 \parallel Lifetime)) > \quad (4.13)$$

In many existing approaches, the *payload* is not encrypted (Vighnesh et al. 2011, Papadimitratos et al. 2006 and Zhang et al. 2008a) when it is communicated with the other vehicles. In order to protect the *payload* (actual data or information) field against eavesdropping and modification by unauthorized users, a protocol is included and explained in section 4.3. To provide two different secure group communications in VANETs, a dual key management scheme is also developed in this research work.

4.3 PROPOSED DUAL KEY MANAGEMENT FOR GROUP COMMUNICATION

Dual Key Management is a group key management scheme in which the TA computes two different group keys intended for two different groups in VANETs. The group is a very important concept in our scheme. Based on the money paid to the TA, a very simple Service Level Agreement (SLA) is considered between the TA and the vehicle users, which categorize the vehicle users into three groups, namely Primary Users (PUs), Secondary Users (SUs) and Unauthorized Users (UUs) in a pre-defined manner. The PUs are eligible to get attractive services such as safety, comfort services and interactive services from the TA. The PUs are authorized VANET users who receive these services from the TA side periodically. The SUs are also authorized VANET users who receive the attractive services such as safety services from the PUs without making any requests to them, but they cannot receive the information directly from the TA. The PUs can communicate with each other by means of V2V communications. However, the SUs can also communicate with each other after getting the SUs group key from the TA through PUs. Both the PUs and the SUs will have a valid VSK received from the TA. Finally, UUs are the vehicle users who do not have access to the information exchanged between PUs and SUs and hence a UU is considered as an intruder in this proposed approach.

To disseminate the information from the TA side to PUs side in a secure way, the TA encrypts the information using a common group key which is derived using individual vehicles secret key of PUs as discussed in one of the previous works (Veltri et al. 2013). Similarly, for broadcasting the information from the PUs to SUs in a secure way, the TA encrypts the group key of SUs using the group key of PUs and multicast it to PUs. All the PUs can get the group key of SUs. This group key is used in the PUs side to encrypt the information and the encrypted message is sent to neighbouring SUs. For computing a common group key separately for PUs and SUs vehicles in the TA side, CRT based group key management scheme is used in this research work.

Let $k_1, k_2, k_3 \ldots k_n$ be pairwise relatively prime positive integers, and let $a_1, a_2, a_3 \ldots a_n$ be positive integers. Then, CRT states that the pair of congruences, $X \equiv a_1 \bmod k_1, X \equiv a_2 \bmod k_2, \ldots, X \equiv a_n \bmod k_n$ has a unique solution $\bmod\ \partial_g = \prod_{i=1}^{n}(k_i)$. To compute the unique solution, the group centre can compute the value as shown in Equation (4.14).

$$X = \sum_{i=1}^{n} a_i \beta_i \gamma_i (mod\ k_i). \qquad (4.14)$$

Where, $\beta_i = \frac{\partial_g}{k_i}$ and $\beta_i \gamma_i \equiv 1\ mod\ k_i$.

The proposed dual group key management scheme works in four phases. The first phase is the TA Initial set up, where a multiplicative group is created at the TA side from which secret key and group key values are selected. For differentiating the VSK values of PUs and SUs vehicles, two types of notations for representing the secret key values used for PUs and Sus are used in this section. The secret key value of PUs is denoted as $PUSK_i$ $(i = 1 \ldots n)$ and SUs are denoted as $SUSK_i$ $(i = 1 \ldots n)$. The second phase is called registration and group key computation phase, where the PUs and SUs complete the registration process and receives $PUSK_i$ and $SUSK_i$ $(i = 1 \ldots n)$ from the TA side. After that, the TA also generates two group keys separately for two groups of PUs and SUs and it informs this group key to them in a secure way. The third phase is secure data transmission, where the data are disseminated using the group key values in the VANET. The final phase of this algorithm is the key updating phase where a group key is updated when an existing PU leaves the PU's multicast group or a new PU joins the PU's multicast group in order to provide forward and backward secrecy. Similarly, the TA also updates the group key of SUs separately.

4.3.1 TA Initial Set Up

Initially, the TA selects large prime numbers p and q, where $p > q$ and $q \leq \left\lceil p/4 \right\rceil$ where p value is used for defining a multiplicative group z_p^* and q is used for selecting the group key values. Initially, the TA selects $PUSK_i$ and $SUSK_i$ from the multiplicative group z_p^* for 'n' number of vehicles which will be given to the vehicle users at the time of offline registration. In the proposed group communication scheme, it is required that all the $PUSK_i$ and $SUSK_i$ values are pairwise relatively prime positive integers and are selected from z_p^* as explained in Zheng et al. (2007) and Zhou & Ou (2009). Moreover, all the secret keys should be much larger than the group key which is selected within the threshold value fixed by q. Next, the TA executes the following steps for computing the group key used for PUs. Similarly, the TA will also compute a group key for SUs.

1) Compute $\partial_g = \prod_{i=1}^{n}(PUSK_i)$. (4.15)

2) Compute $x_i = \frac{\partial_g}{PUSK_i}$ where $i = 1,2,3 \dots n$ (4.16)

3) Compute y_i such that $x_i \times y_i \equiv 1 \, mod \, PUSK_i$ (4.17)

4) Multiply all users x_i and y_i values and store them in the variables,

$$var_i = x_i \times y_i \qquad (4.18)$$

5) Compute the value $\mu = \sum_i^n var_i \qquad (4.19)$

4.3.2 Group Key Computation

In this phase, the VANET group users complete the registration process and get their corresponding group secret keys from the TA. Whenever the TA wants to send common information to a group of VANET users (PUs) to support the group communication, the TA computes the group key in the following way and multicast it to the PUs group through RSU.

a) Initially, the TA selects a random element k_{pug} as a new group key for PUs within the range q.

b) Multiply the newly generated group key with the value μ which is computed in TA initial setup.

$$\gamma_{pug} = k_{pug} \times \mu \qquad (4.20)$$

c) The TA broadcast a single message γ_{pug} to the VANET users. Upon receiving γ_{pug} value from the TA side, an authorized vehicle can obtain the new group key k_{pug} by doing only one modulo division operation as shown in Equation (4.21).

$$\gamma_{pug} \bmod PUSK_i = k_{pug} \qquad (4.21)$$

Since, $k_{pug} < q < PUSK_i < p$ and $\mu \bmod PUSK_i = 1$, the k_{pug} obtained in this way must be equal to the k_{pug} generated in

Step a) of group key computation phase. After computing the group key, the TA also computes another group key k_{sug} using the aforementioned procedure for SUs. Then, it encrypts this k_{sug} using k_{pug} and it is sent as a multicast message along with γ_{pug} and γ_{sug} to all the PUs.

$$< E_{k_{pug}}(k_{sug}) \parallel \gamma_{pug} \parallel \gamma_{sug} > \quad (4.22)$$

After receiving the packet from TA, the PUs compute the value of k_{pug} from γ_{pug} using Equation (4.21) and then decrypt $E_{k_{pug}}(k_{sug})$ to get the group key value of SUs.

$$< D_{k_{pug}}\left(E_{k_{pug}}(k_{sug})\right) \parallel \gamma_{sug} > \quad (4.23)$$

Then the PUs send γ_{sug} as a multicast message to all the SUs in its coverage area. After receiving this message from the PUs, the SUs compute the value of k_{sug} from γ_{sug} as given in Equation (4.24).

$$\gamma_{sug} \bmod SUSK_i = k_{sug} \quad (4.24)$$

The PUs utilize the group key value of SUs to broadcast the information to the nearest SUs within their coverage area. Therefore, the TA encrypts the information using this group key (k_{pug}) and multicast it to the PUs. All the PUs can use their group key to decrypt the information received from the TA side. Each PU can in turn broadcast the information received from the TA to SUs by encrypting it using k_{sug}. In this way, the secure group communication is implemented in this proposed work. When 'i' reaches to n, the TA executes TA Initial set up phase to compute ∂_g, var_i and μ for 'm'

number of users where $m = n \times \delta$. The value δ is a constant value which may take values less than 5 depending upon the dynamic nature of the multicast group.

4.3.3 Secure Data Transmission in VANETs

This subsection explains the secure transmission of data (information) from TA to vehicles and between vehicles in VANETs. Figure 4.2 shows the working of secure data transmission that takes place between the TA and PUs. In addition to this, it also represents the V2V communications that take place between PUs and SUs. The TA has collection of servers for storing the necessary keys and data required for the VANET users. The TA can multicast the information to PUs through a dedicated Internet connection. The PUs in turn can broadcast the information to SUs with the PUs wireless medium. Finally, the UUs have no permission to communicate to the VANETs since they are unauthorized users. In order to improve the confidentiality, the messages should be exchanged in an encrypted form so that the UUs cannot access the messages. The steps involved in the secure data transmission in VANET communication are described as follows:

Step 1. The TA generates a group key (k_{pug}) using the PUSK's of the PUs after collecting the requests of PUs through any RSU. Also, it generates a separate group key (k_{sug}) for SUs.

Step 2. Then, the TA multicasts both the group key values in an encrypted form as explained in Equation (4.20). Both the group users can find their group key using their secret key values as used in Equation (4.21). Also, the TA sends the group key value of SUs through the RSU to the PUs by encrypting it using PUs group key $E_{k_{pug}}(k_{sug})$.

Step 3. The TA sends messages or traffic information to the PUs only by encrypting the messages using PUs group key k_{pug}, and there is no message exchange between the TA and SUs.

$$< E_{k_{pug}}(ID_{TA} \parallel message\ or\ information) > \qquad (4.25)$$

Step 4. After receiving the data packet from the TA, the PUs decrypt the packet using k_{pug} and consumes the information or messages.

$$< D_{k_{pug}}(E_{k_{pug}}(ID_{TA} \parallel message\ or\ information)) > \qquad (4.26)$$

Step 5. The PUs broadcast the message to the SUs, by encrypting the message or information using the group key of SUs along with the authentication code received from the TA in the dual authentication technique.

$$< E_{k_{sug}}(ID_V \parallel messge\ or\ information\ ) \parallel E_{TA-Pvt}(AC \parallel ID_V \parallel TS_3 \parallel Lifetime) > \qquad (4.27)$$

Step 6. After receiving the data packet, the nearest SUs can decrypt the data packet using the group key k_{sug} and can also verify the authenticity of the messages by decrypting the authentication part using the public key $(TA - Pub)$ of TA as shown below:

$$< D_{TA-Pub}\big(E_{TA-Pvt}(AC \parallel ID_V \parallel TS_3 \parallel Lifetime)\big) > \qquad (4.28)$$

Step 7. The SUs can in turn forward the received data packet to other SUs by encrypting it using k_{sug} over a long range using multihop communication.

Step 8. After receiving the packets, the SUs can decrypt the packet using the k_{sug} and process the messages.

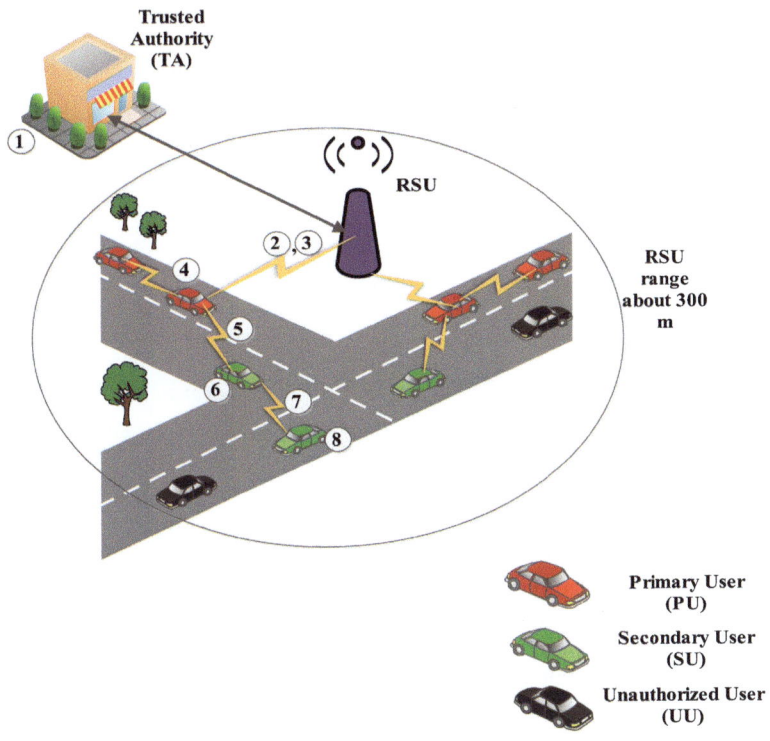

Figure 4.2 Secure data communication in VANET

4.3.4 Key Updating

Group key updating operation is performed when a PU joins or leaves and usually takes more computational complexity in most of the group key management schemes (Papadimitratos et al. 2006, Veltri et al. 2013 and Sun et al. 2007). When a PU joins the VANET group, it is the responsibility of the TA to communicate the new group key in a secure way to the group members. Therefore, the newly joining user cannot view the previous communications and it provides backward secrecy. Similarly, when a PU leaves from a group, the TA must update the group key in order to avoid the use of a new group key by the old PU to preserve forward secrecy. In our proposed key management scheme, the group key updating process is performed in a simplest way when the group membership changes. For example, when a vehicle v_i of PU leaves the group, the TA has to perform the following steps.

1)   Subtract var_i from μ.

$$\mu' = \mu - var_i \qquad (4.29)$$

2)   Next, the TA must select a new group key k'_{pug} and it should be multiplied by μ' to form the rekeying message as shown below.

$$\gamma_{pug}' = k'_{pug} \times \mu' \qquad (4.30)$$

3)   The updated group key value is sent as a broadcast message to all the existing PUs. The existing users of the PUs group can get the updated group key value k'_{pug} by doing only one mod operation as shown in Equation (4.21). From the received value, the vehicle v_i cannot find the newly updated group key k'_{pug} since that particular vehicle's secret key is not included in μ'.

Similarly, if a PU wants to join in the multicast group, then the TA has to perform only one addition operation for updating the group key. For example, if v_i wants to join an existing VANET group, then the TA has to perform the following steps for group key updating.

1) Instead of computing x_i and y_i value for the new VANET user, the TA can take the multiplied value of x_i and y_i from the variable var_i which is already computed in the TA initialization phase. The TA can select this value from the TA's storage area to compute $\mu' = \mu + var_i$.

2) Next, the TA selects a new group key k'_{pug} and multiplies it with updated μ' to form the rekeying message as shown in Equation (4.30).

3) The updated group key value is sent as a multicast message to all the existing and newly joined PUs of the group. From the multicast value γ_{pug}', the newly joined PUs of the multicast group can find the newly updated group key k'_{pug} since his/her var_i value is included in μ' using var_i.

Therefore, in general, if $'n'$ PUs want to join in the existing PU's multicast group, the TA has to perform $'n'$ additions for updating the group key. The key strength of this proposed algorithm is that the computational complexity of the TA is completely reduced in comparison to the other existing approaches (Zhang et al. 2008a and Veltri et al. 2013). The computation complexity of the TA is $O(1)$ when a single PU joins or leaves from the multicast group. In addition to this, the computational complexity of a multicast PU is also minimized by allowing each PU to perform only one modulo division operation. Moreover, the TA takes only one broadcast message which is same in most of the existing algorithms for informing the updated group key value to PUs of the multicast group.

4.4 SECURITY ANALYSIS

In this section, the security strength of the proposed dual authentication scheme is analysed. In addition to that, the proposed group key management scheme is analyzed for various attacks to support forward secrecy and backward secrecy as discussed in many existing algorithms (Wong et al. 2000, Naranjo et al. 2012, Vijayakumar et al. 2013 and Veltri et al. 2013). The assumption of the implemented key management scheme is that an adversary might be a PU for some time and the TA keeps all user secret keys secretly.

4.4.1 Resistance to Replay Attack

In a replay attack, the malicious user re-injects the previously received messages or packets back into the VANET. To protect our system from replay attack and provide freshness to messages, our proposed scheme maintains time stamps to keep a cache of recently received messages through which the newly received messages can be compared.

4.4.2 Masquerade and Sybil Attacks

In this section, the security properties of the proposed dual authentication scheme shows how the scheme is effective for resisting masquerade and Sybil attacks. In many existing approaches, TA is unable to distinguish an authentication effort from a malicious attacker. Because, the malicious attacker makes use of real users' authentication efforts with stolen passwords, usernames and secret keys, and the TA still considers the attackers as real users. In this research work, a novel dual authentication scheme has been proposed, which can effectively oppose the malicious behaviour of the attackers that is previously mentioned. In the proposed authentication scheme, even if the attacker knows the *VSK* of any vehicle user, the OBU verifies the fingerprint of the vehicle user. If it doesn't match, the particular vehicle is not allowed to make communication with VANETs. Hence, the masquerade and Sybil attacks are successfully prevented in the proposed dual authentication scheme.

4.4.3 Message Tampering /Fabrication/Alteration Attack

In this research work, the messages are encrypted using the group keys in the group communication before they are sent among the groups. For example, the TA sends messages to the PUs group by encrypting the messages using the PUs group key k_{pug}. Therefore, no one can delete, modify and alter the content of the messages during the transmission between the TA and PUs. Since, the group keys are managed by the TA, an intruder will not be able to find the key in a feasible amount of time to communicate with the group.

4.4.4 Backward Secrecy

Backward secrecy is the technique of preventing a new PU from accessing the previous communication before joining the group. In order to access the previous communication, an adversary needs to obtain the previous group key. Moreover, if the adversary becomes a PU in a group, it may try to derive the previous group key which is not permitted. In the proposed group key management scheme, when the newly updated group key is communicated to old group members, an adversary needs to find any one of the PUs secret key. Moreover, all the $PUSK$'s are randomly selected from a large set of positive integers with respect to the multiplicative group. Even if the adversary finds any one of the PUs secret key $PUSK_i$, then the adversary cannot use this $PUSK_i$. Because, the dual authentication scheme is used in this proposed approach to participate in VANET communication. When the adversary tries to use any other PUs $PUSK_i$, the TA will also ask the adversary user to complete the authentication process to get authentication code before participating in the VANET's group communication. Moreover, if an adversary sends any information without including the authentication code, then the receiving vehicles will not process the information. This property makes the situation infeasible for the adversary to use any other PUs secret key. Consequently, the adversary cannot access the communication sent before join, which means the proposed approach supports the initial security requirement.

4.4.5 Forward Secrecy

Forward secrecy is the technique of preventing a PU from accessing current communication after leave operation. When a PU leaves the group, he or she may try to derive the group key by using any attacking methods. In the proposed algorithm, it is infeasible for a PU to compute the current group key after the leave operation from the group that was explained for the backward secrecy technique. Because, when a PU v_i leaves from the group, the TA subtract his or her share value such as multiplication of x_i and y_i which is stored in var_i from μ value to produce μ'. This updated μ' is multiplied by the newly generated group key value k'_{pug} to form the rekeying message γ_{pug}'. Therefore, a PU who had already left for the service cannot find the new group key in a feasible way since his or her personal keying information is not included. The PU who had left from the group may try to find k'_{pug} from the rekeying value which is sent as a broadcast message from the TA in an infeasible method. In order to do that, the PU has to multiply his or her secret key value with all the numbers starting from 1 to q where q is the maximum limit of group key value. At a certain point, it will give a value $\vartheta = \gamma_{pug}'$ (i.e. $PUSK_i \times \omega = \vartheta$). After finding this ω value, the PU v_i can find a set of numbers S that will divide the number ω. Therefore, the value of S is defined as the set of numbers $\{\omega \bmod 1, \omega \bmod 2, ..., \omega \bmod \omega\} = 0$. Among the set of numbers, newly generated group key k'_{pug} is also one of the number $(i, e., k'_{pug} \in S)$. In this case, if the size of $PUSK_i$ is w bits, then the attacker has to perform 2^w multiplication. The time taken to derive k'_{pug} can be increased by choosing a large $PUSK_i$ for each VANET user's secret key. In this work, the size of $PUSK_i$ must be 1024 bits and prior experiments were conducted with 128 bits, 256 bits and 512 bits. After finding the set of values S that divides the number ω, the attacker (user left from the group) can find the new group key by selecting the values from the set S by using brute force attack by making 2^{s-1} attempts. Consequently, an adversary cannot find the group key in a feasible method in order to access the current communication, which means the second security requirement is also supported in our proposed algorithm.

4.4.6 Collusion Attack

The Collusion attack is the one in which two or more adversaries act as legitimate PUs when they are participating in the group and then cooperatively compute the updated group key after leaving the group. Since, the value of var_i is subtracted from μ after the leaving operation is performed in a multicast group, any number of prior user's collision will not be used to gain information about the congruence system and to derive the updated group key k'_{pug} as long as the pairwise relatively prime numbers are large. The following scenario describes a kind of collusion attack in which two adversaries act as legitimate users. Consider v_1 as an adversary A who knows the key values $PUSK_1$, k_{pug} and v_3 as an adversary B who knows the key values $PUSK_3$ and k_{pug} at time $'t-2'$. In time $'t-1'$, the adversary A leaves the group with the key values $PUSK_1$ and k_{pug}. B receives the rekeying message γ_{pug}' from the TA at the time $'t'$ and computes k'_{pug}. In time $'t+1'$, B leaves the group with the two key values $PUSK_3$ and k'_{pug}. Both of these adversaries exchange their known key values $PUSK_1, k_{pug}, PUSK_3$ and k'_{pug}. Using these known values, the adversaries A and B cannot cooperatively find the updated group key k'_{pug} which is broadcast at time $'t+2'$ in a feasible amount of time since their shares var_1 and var_3 are excluded from μ.

4.5 PERFORMANCE ANALYSIS

The proposed research work is analysed in terms of two performance metrics namely the computation time and communication time for updating the group key in order to perform secure group communication in the PUs of VANET communication. The computation time is defined as the time taken to compute group key at the TA when group membership changes in the VANET group. The communication time is defined as the time taken to broadcast the amount of information from TA in order to make the VANET users to recover the group key. Table 4.1 shows the computation and storage complexities of various key management approaches, namely Chinese Remainder Group Key (CRGK) (Zheng et al. 2007), Fast-chinese Remainder Group Key (FRGK) (Syamsuddin et al. 2008), Key-tree Chinese Remainder Theorem (KCRT) (Zhou & Ou 2009), Number Theory Research Unit (NTRU) (Lv et al. 2012) and Elgamal Group Key Management (EGKM) (Lv et al. 2012) and the proposed VANET Group Key Management (VGKM) which are based on the CRT. The notations used for comparisons are defined as: n is the number of users, τ is the maximum number of children of each node of the tree, EEA is the time taken to find the inverse element of a multiplicative group using Extended Euclidean Algorithm, exp represtens the exponential operation, M represents the multiplication operation, D represents the division operation, A represents the addition operation and S represents the subtraction operation.

Among these schemes, the Number Theory Research Unit (NTRU) based group key management scheme uses a multiplication ring from which it chooses some polynomial values as private and public keys from which it computes a common group key. Hence, the multiplication operation used in this scheme is performed by using the convolution product method. All the remaining schemes use a multiplicative group for choosing and computing the keys. Moreover, all the existing schemes take $O(n)$ for updating the group key when a single AV user joins or leaves from the secure VANET communication. From Table 4.1, it is evident that all the existing approaches take more computation complexity if it is used in the TA side in the VANET for computing the group key for performing a single user join/leave operation which is very high in comparison with the proposed approach. Therefore, the proposed approach takes less computation complexity when it is compared with all the remaining five approaches since it takes only 1 subtraction operation or (addition) operation to be performed when a single user leave or join operation is performed. Moreover, the proposed approach doesn't perform any cyclic convolution product operation and multiplicative inverse operation on the user side which reduces user's computational complexity. The amount of information bits necessary to be communicated while updating the group key to our proposed approach and existing approaches are calculated and are also shown in Table 4.1. It is very clear that the proposed group key management scheme takes the same communication complexity as that of most of the existing group key management protocol which are based on CRT

Table 4.1 Computation, storage and communication complexities of various schemes

Parameters	CRGK	FRGK	KCRT	NTRU	EGKM	VGKM
Computation Cost (TA)	$O(n)$ (xor + A + M + EEA)	$O(n)$ (xor + A + M)	$O(\log_\tau n)$ (xor + A + M + EEA)	$O(n)(M + A + D + EEA)$	$O(n)(M + A + D + EEA)$	$O(1)$ (A or S)
Computation Cost (User)	1mod + 1xor	1mod + 1xor	1mod + 1xor	(2M + 1A + 1mod + 1EEA)	(1M + 1exp + 1mod + 1EEA)	1mod
Storage Complexity (user)	2	2	$(\log_\tau n)$	4	3	2
Storage Complexity (TA)	$2n + 1$	$4n + 1$	$2n - 1$	$2n + 7$	$2n + 5$	$4n + 3$
Communication Complexity	1 broadcast	1 broadcast	1 broadcast	n	n	1 broadcast

The proposed method has been executed in JAVA (Intel Core i3 processor, 2GB RAM, 500 GB Hard disk, Windows XP Operating System) for a group of 1000 nodes and each node is considered as a VANET user. For implementing this authenticated group key management scheme suitable for VANET, the TA generates $PUSK_i$ values for 1000 nodes randomly. The $PUSK_i$ values used in this approach are 1024 bit positive integers which are relatively prime. For generating large integers, the Big integer class is used which supports various methods for handling large positive integers. The method multiply() supported by Big integer class is used to multiply all users secret key into a variable which will be used to find x_i and y_i values. The method modInverse() is used to find the multiplicative inverse of a given element with respect to the size of the multiplicative group. The proposed group key computation scheme takes less computational complexity because it takes only addition or subtraction operation in the key updating process. Moreover, for computing the group key in all the existing approaches present in the literature, the computation time is measured separately for x_i which is obtained by dividing ∂_g and y_i which is obtained by finding the multiplicative inverse for x_i. All the existing algorithms shown in Table 4.1 takes more computational time for calculating x_i and y_i values, which would increase the computing load of the TA in VANETs. In the proposed approach, computational complexity is very much reduced because 1) calculating x_i and y_i value is neglected by storing them in in the TA's server storage area and 2) multiplying x_i with y_i is also reduced, which is done in the TA initialization phase. Therefore, the proposed VGKM approach reduces the computing load of TA by slightly increasing the storage overhead of the TA.

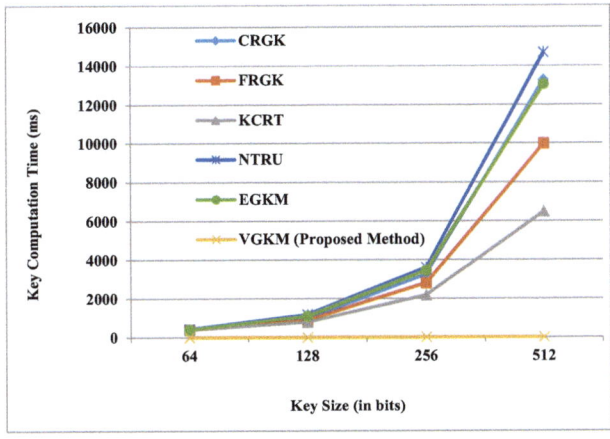

Figure 4.3 Group key computation time at TA side

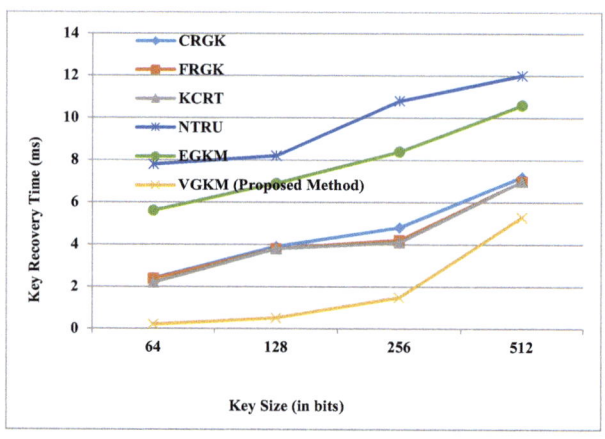

Figure 4.4 PUs key recovery time in the VANET

The graphical results shown in Figure 4.3 are used to compare the group key computation time of TA for the proposed method with the existing methods. It compares the results obtained from the proposed VGKM with CRGK, FRGK, KCRT, NTRU and EGKM. From Figure 4.3, it is observed that when the key is 512 bits, the group key computation time of TA is found to be 19ms in the proposed approach, which is better in comparison with the other existing schemes. The results shown in Figure 4.4 are used to compare the PUs key recovery time of the proposed method with the existing methods. It compares the results obtained from the proposed scheme with existing approaches and it is observed that when the key size is 512 bits, the key recovery time of a user is found to be 5.3ms in the proposed approach, which is better in comparison with the other existing schemes.

4.6 CONCLUSIONS

In this research work, a new dual authentication scheme is proposed for improving the security of vehicles that are communicating with the VANET environment. For providing such authentication in dual mode, two components such as hash code and fingerprint of each communicating vehicle user are used. Therefore, the fingerprint authentication technique is integrated into a hash code creation method in this research to avoid malicious users to use the secret key of any VANET users in order to participate in the VANET communication. Moreover, to avoid malicious users from spoofing the authentication code issued for any VANET users and sending erroneous messages to other vehicles a new dual key management scheme is introduced in this research work. The dual key management scheme implemented in this research work is computationally efficient that supports secure data transmission from TA to PUs and PUs to SUs based on two different group keys, one for PUs and another one for SUs for further improving the security among different classes of vehicles. Moreover, the proposed algorithm also takes single broadcast messages from TA to inform the group members in order to recover the updated group key.

CHAPTER 5

CPAV: COMPUTATIONALLY EFFICIENT PRIVACY PRESERVING ANONYMOUS AUTHENTICATION FOR A VEHICLE USER IN VANETS

5.1 INTRODUCTION

With the progression and wide use of wireless communication technologies, vehicular ad hoc networks (VANETs) will be expected as a hopeful approach to provide safety related and infotainment services to vehicle users. Since, the V2V and V2R communications are performed over the open wireless channel (Oh et al. 1999), it is necessary to preserve the privacy of vehicle users during the authentication process with very low computational cost. Moreover, to prevent malicious users and malfunctioning OBUs continuing with the system, an efficient conditional tracking mechanism is also required for VANETs. Hence, in this research work, a computationally efficient anonymous authentication scheme is proposed to preserve the privacy of vehicle users during the authentication. Moreover, conditional privacy preservation scheme is proposed to avoid the misbehavouring users continuing with the VANET system.

5.2 SECURITY REQUIREMENTS

The primary goal of designing this research work is to provide a computationally efficient conditional privacy preserving anonymous authentication to satisfy the following security requirements:

1) **Message integrity and authentication**: When a vehicle enters into the region of an RSU, it should be authenticated by the RSU in an anonymous manner before it issues the safety related messages. On the other hand, a vehicle should authenticate other vehicles before it receives a message from it in an anonymous manner. Moreover, each message that is sent in the VANET system is appended with an anonymous signature to preserve the integrity of the transmitted message.

2) **Identity privacy preserving:** The real identity of each vehicle should be kept secret from other entities in the VANET system to preserve vehicle's privacy from various attacks.

3) **Traceability:** Even if a vehicle's real identity is kept hidden from entities in the network, the TA has the capability to get a vehicle's real identity from its anonymous certificate in the case of dispute or misbehaviour.

5.3 BILINEAR PAIRING

Let G_1, G_2, and G_T denote three multiplicative cyclic groups of order q, where q is a large prime. Suppose that G_1, G_2, and G_T are equipped with pairing. The bilinear map $e : G_1 \times G_2 \to G_T$ satisfies three properties.

1. **Bilinearity:** The mapping $e : G_1 \times G_2 \to G_T$ is said to be bilinear if $e(g_1^a, g_2^b) = e(g_1, g_2)^{ab}, g_1 \in G_1 \ \& \ g_2 \in G_2$ and $\forall a, b \in Z_q^*$, where $Z_q^* = [1, \ldots, q-1]$.

2. **Nondegeneracy:** $e(g_1, g_2) \neq 1_{G_T}$.

3. **Computability:** There exists an efficient algorithm to easily compute the bilinear map $e : G_1 \times G_2 \to G_T$.

The isomorphism is denoted by ψ and hence a well computable isomorphism $\psi : G_2 \rightarrow G_1$ is basically required. The group that possesses such a map e is called a bilinear group.

5.4 PROPOSED CPAV SCHEME

The proposed scheme has four phases namely system initialization phase, registration phase, secure activation key distribution phase and CPAV secure anonymous mutual authentication phase. In order to confirm the integrity and authentication of the messages, the CPAV scheme uses bilinear pairing technique (Boneh et al. 2004), which is considered as the basis of this scheme.

5.4.1 System Initialization

The TA first selects two random numbers $m, n \in Z_q^*$, which are kept secret to effectively compute the public keys from the private keys. Moreover, the TA selects a secure cryptographic hash function: $H: \{0,1\}^* \rightarrow Z_q^*$. The TA also selects a private key $prk = u \in Z_q^*$ and computes its corresponding public key $puk = g_1^{u+n}$. Once the vehicle V_j has submitted the necessary documents to TA, it chooses a user private key $uprk = v \in Z_q^*$ and computes the corresponding user public key $upuk = g_1^{v+m}$. Therefore, it generates four key values namely private key, public key, user private key and the user public key for various computation purposes. After computing these values, the TA publishes the system parameters as $param = (q, G_1, G_2, G_T, e, g_1, g_2, puk, H)$.

5.4.2 Registration

1. First, V_j is required to register his/her Tamper Proof Device (TPD) to the TA. After this registration, the TA assigns a password $(pw) = g_1^{m+n}$ and also computes the activation key $s = g_1^{v+n+m+u}$ for TPD. Then, TA computes a re-encryption key $REK = pw * upuk$.

2. TA then generates the VANET license (VL_{V_j}) for each vehicle V_j, where $VL_{V_j} = upuk^m * g_1^m$.

3. TA preloads $uprk$ and $upuk$ in vehicles TPD and provides pw, VL_{V_j} and REK to the vehicle user after the successful completion of his/her offline registration.

4. Next, TA keeps $(ID - V_j, upuk^{m*n})$ in its tracking list, where $ID - V_j$ is the identity (ID) of V_j assigned by the TA at the time of its registration.

5.4.3 Secure Activation Key Distribution

When a vehicle V_j is started to travel into the road, it is required to activate the TPD to get the $uprk, upuk$ for making anonymous communication with other entities in the VANET system. To get the TPD activation key, V_j sends its licence by encrypting (E) it using TA's public key. Hence, V_j sends $E_{puk}(VL_{V_j})$ to TA through an RSU that it first meets. Let us consider RSU_x is the first RSU that a vehicle V_j meets first. After receiving it, the TA decrypting it using its private key $D_{prk}(E_{puk}(VL_{V_j}))$ and computes the secret message SM as given below

$$SM = s * pw * upuk$$

Then, the TA sends $E_{pw}(SM)$ through RSU_x to V_j. After receiving this, V_j computes, $D_{pw}(E_{pw}(SM))$ and get the SM. From the value of SM, V_j extracts the TPD activation key as given below:

$$activation\ key = \frac{SM}{REK} = \frac{s * pw * upuk}{pw * upuk} = s$$

If either the password or the activation key, or both are incorrect, then it is not possible to activate the TPD to get $uprk$ and $upuk$. If both the password and activation key are correct, then the TPD will give $uprk$ and $upuk$ for anonymous communication.

5.4.4 CPAV Secure Anonymous Mutual Authentication

When a vehicle V_j enters into the region of RSU_x or it initiates communication with other vehicles, the mutual authentication between the vehicle and the RSU or other vehicles should be performed to avoid communication with malicious vehicles. In order to preserve its privacy from the RSUs and other vehicles, the vehicle V_j use short life anonymous keys for authentication. Once the vehicle user leaves from the region of the RSU_x and enter into the domain of another RSU, it requires a new authentication process.

1) The vehicle V_j first chooses a random number x_j from a set of k random numbers $x_1, x_2, \ldots x_k \in Z_N^*$ as the short-time private keys and then computes the corresponding public keys $y_j = g_1^{x_j + uprk}$ for $j = 1, 2, \ldots, k$.

2) For each short-time public key y_j, the vehicle user computes the short-time anonymous self-generated certificate cer_j as follows:

   a) The vehicle user randomly choose $h_1 \in Z_q^*$ and compute s_1 and s_2 where

   $$s_1 = g_1^{uprk}, s_2 = g_1^{uprk+h_1}$$

   b) Then, computes the challenger $C = H(y_j \| s_1 \| s_2)$ as well as s_1' and s_2' where

   $$s_1' = g_1^{x_j-h_1}, \quad s_2' = \frac{1}{g_1^{x_j}}$$

   c) Finally, set $cer_j = \{y_j \| s_1' \| s_2' \| C\}$ as the short-time anonymous self-generated certificate.

3) Then, the vehicle user can use one short-time x_j, y_j and cer_j for anonymous authentication. In order to preserve the integrity of a message M, the vehicle user computes the signature $\sigma = g_2^{\frac{1}{x_j+uprk+H(M)}}$ and broadcast the following value to all other vehicles.

$$msg = (M \| \sigma \| y_j \| cer_j \| VL_{v_j})$$

4) After receiving this message, everyone can verify the validity of the message source and integrity of the message as follows. The receiver first computes

$$S_a = y_j \times s_2'$$

83

$$s_b = \frac{y_j}{s_1'}$$

and then verifies whether $C = H(y_j \parallel s_a \parallel s_b)$. If it holds, the y_j and cer_j passes the verification and the vehicle user is authenticated by the receiver successfully.

Proof of correctness:

$$s_a = y_j \times s_2' = g_1^{x_j + uprk} \times \frac{1}{g_1^{x_j}}$$

$$= g_1^{x_j + uprk - x_j} = g_1^{uprk} = s_1$$

$$s_b = \frac{y_j}{s_1'} = \frac{g_1^{x_j + uprk}}{g_1^{x_j - h_1}}$$

$$= g_1^{x_j + uprk - x_j + h_1} = g_1^{uprk + h_1} = s_2$$

5) Once the verifier has checked the certificate, the verifier checks the integrity of M as follows:

$$e(y_j \cdot g_1^{H(M)}, \sigma) = e(g_1, g_2)$$

If it holds, then the verifier accepts the message M, otherwise it will be rejected.

Proof of correctness:

$$e(y_j \cdot g_1^{H(M)}, \sigma) = e\left(g_1^{x_j + uprk} \cdot g_1^{H(M)}, g_2^{\frac{1}{x_j + uprk + H(M)}}\right)$$

$$= e\left(g_1^{x_j+uprk+H(M)}, g_2^{\frac{1}{x_j+uprk+H(M)}}\right)$$

$$= e(g_1, g_2) \text{ (By bilinear property)}$$

Here, the value of $e(g1, g2)$ can be pre-computed.

Conditional Tracking: If the accepted message M under the vehicle license VL_{V_j} has been disputed, then the TA can efficiently trace the real identity $ID - V_j$, by looking up the entry $(ID - V_j, upuk^{m*n})$ in the tracking list.

$$\frac{\left(VL_{V_j}\right)^n}{g_1^{m*n}} = \frac{(upuk^m * g_1^m)^n}{g_1^{m*n}} = \frac{upuk^{m*n} * g_1^{m*n}}{g_1^{m*n}} = upuk^{m*n}$$

5.5   SECURITY ANALYSIS

In this section, the proposed protocol is analysed with some security and privacy issues. In CPAV, to perform an impersonation attack, the attacker should derive the temporary short time keys owned by a legitimate vehicle and the user private key issued by the TA to a particular vehicle. However, the attacker cannot compromise the registration protocol because it is performed in offline mode directly in the TA. Hence, the proposed research work is semantically secure against impersonation attack.

5.5.1   Message Integrity and Source Authentication

Generally, the message integrity can be ensured by the signature attached with each message. In the anonymous secure mutual authentication, the signature on message M is defined as $\sigma = g_2^{\frac{1}{x_j+uprk+H(M)}}$. In this signature, the temporary short time private key x_j and the user private key $uprk$ are only known to the particular vehicle. So, no other users can forge the signature.

Moreover, x_j value is changed periodically. Therefore, even if a temporary short time private key x_j is found, it is infeasible to forge the signature. Similarly, the vehicle certificates are generated using the vehicle's private key $uprk$ and the short-time private key x_j. Hence, no other user can forge the certificate. When a vehicle V_j enters into the region of RSU_x, the messages communicated between RSU_x and V_j contain the signature and certificate, can guarantee the message integrity and source authentication. Therefore, impersonation attack (Gamage et al. 2006) can be avoided due to the nature of message integrity and source authentication.

5.5.2 Conditional Privacy Preservation

In CPAV authentication, vehicles and RSU's use anonymous certificates and signatures to protect their real identities from other users. However, the TA has the capability to trace the real identity of a vehicle or an RSU from its anonymous certificate. For instance, when a vehicle sends a bogus message along with an anonymous certificate, the TA can check the content of the message. If it is bogus, then the TA gets the anonymous certificate of that message and maps the anonymous certificate with the tracking list.

From the mapping, the TA can trace the real identity of a vehicle effectively. After that, the TA can disclose the privacy of the vehicle user and revoke that user from VANET.

5.5.3 Anonymity

With a valid signature $\sigma = g_2^{\frac{1}{x_j + uprk + H(M)}}$ and certificate $Cert_j$, it is computationally hard to identify the actual sender of the message. Hence, the attacker gets zero knowledge about the sender from σ and $Cert_j$.

5.6 PERFORMANCE ANALYSIS

In this section, the performance of the proposed CPAV authentication is evaluated in terms of computational cost for certificate and signature verification process. The computational cost is defined as the total time required for verifying either one signature and one certificate or n signatures and n certificates to authenticate a vehicle and to check the integrity of a message. The computational cost of CPAV authentication scheme is compared with many existing schemes BLS (Boneh et al. 2003), ECPP (Lu et al. 2008), CAS (Gong et al. 2007), GSB (Lin et al. 2007), KPSD (Lu et al. 2012) Among the various existing schemes, KPSD is a certificate based signature verification system proposed in (Lu et al. 2008) which is the foundation of the proposed CPAV verification scheme. Let T_p is the time required for performing a pairing operation, T_h is the time required for performing a hash operation and the time required for performing one multiplication is T_m. The time needed to perform exponentiation operation in G_1 and G_2 are denoted as T_{ep-1} and T_{ep-2}.

From Table 5.1, it can be observed that the proposed CPAV scheme takes low computational cost among the various existing schemes to perform certificate and signature verification process. Because, the CPAV scheme takes only $2T_p, 2T_m$ and T_h for verifying one certificate & signature. Therefore, the proposed CPAV scheme can verify maximum numbers of signatures and certificates within 300ms compared to BLS, ECPP, CAS, GSB and KPSD schemes. It can be seen that T_p and T_h are the most time-consuming operations in the signature verification process. Among the various existing schemes, the proposed CPAV scheme use only two pairing operations for verifying one signature and requires only $(1 + n)$ pairing operations for verifying n signatures. Therefore the proposed CPAV scheme takes less computational cost in comparison with all the existing schemes.

For performing the hash operation, exponential operation, multiplication and pairing operation, the pairing-based cryptography (PBC 2005) library is used in this paper. For the aforementioned operations, the Type-A curve defined in the PBC library is used with the default parameters. In order to measure the actual computation time of the proposed EAAP scheme, a 2-GHz machine with 4-GB installed memory, running Cygwin 1.7.35-15 (Cygwin 2016) with the gcc version 4.9.2 is used for the implementations in this research work.

Table 5.1 Certificate and signature verification cost of various schemes

Method	For one Certificate & Signature	For n Certificates & Signatures
BLS	$4T_p + 2T_h$	$(2n + 2)T_p + 2nT_h$
ECPP	$3T_p + 11T_m + T_h$	$3nT_p + (10 + n)T_m + nT_h$
CAS	$5T_p + 2T_h$	$(4n + 1)T_p + 2nT_h$
GSB	$3T_p + 4T_{ep-1} + 5T_{ep-2} + T_h$	$3nT_p + 4nT_{ep-1} + 5nT_{ep-2} + nT_h$
KPSD	$4T_p + 5T_{ep-1} + 5T_{ep-2} + T_h$	$(3+n)T_p + (4+n)T_{ep-1} + 5nT_{ep-2} + nT_h$
CPAV (Proposed)	$2T_p + 2T_m + T_h$	$(1+n)T_p + 2nT_m + nT_h$

All the results are analysed over 100 randomized simulation runs and then the average of the results is considered. In the simulations, the time parameters T_p, T_h and T_m are measured and it is found to be equal to 1.6 ms (milliseconds), 2.7 ms, and 0.001 ms, respectively. The time needed to perform exponentiation T_{ep-1} and T_{ep-2} are found to be equal to 0.7 ms and 0.6 ms respectively. A certificate which has been sent with a message signature has been taken for computing the verification time. In Table 5.1, the certificate and signature verification cost for BLS, ECPP, CAS, GSB, KPSD and CPAV schemes are summarized.

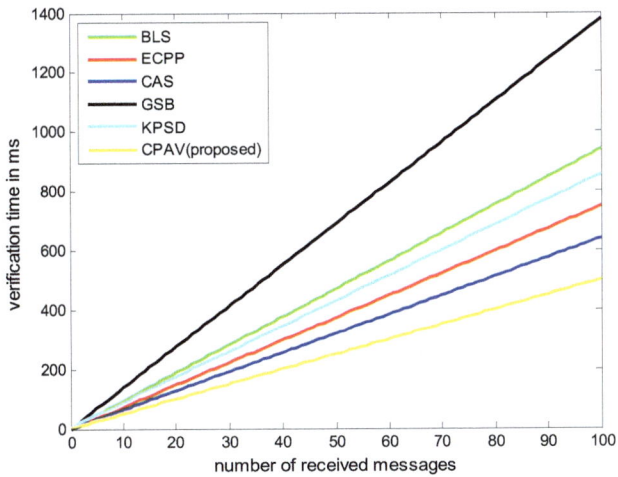

Figure 5.1 Comparison of certificate and signature verification time

Figure 5.1 clearly shows the verification time in ms for the number of the received messages (n). It can be seen that when n is large, the proposed CPAV authentication scheme is much more efficient than the other existing schemes and affords the lowest verification time among the schemes under comparison. From Figure 5.1, it is very clear to understand that the proposed CPAV authentication scheme takes only 420 ms for verifying 100 certificates & signatures. However, other existing schemes take more than 600 ms for verifying 100 certificates & signatures.

In order to evaluate the metric of total computational delay of anonymous authentication in CPAV, the computational delays caused by certificate and signature generation by the sending vehicle and verification during authentication by the receiving vehicle are considered. For CPAV, the total computational cost (TCC) of the anonymous authentication T_{TCC} is calculated as:

$$T_{TCC} = T_{gen}^{sender} + T_{verify}^{receiver}$$

Where T_{gen}^{sender} denotes the time required to generate one anonymous certificate and signature by the sending vehicle, and $T_{verify}^{receiver}$ denotes the time required to verify one anonymous certificate and signature by the receiving vehicle, by using the CPAV scheme. In the proposed research work, the TCC of CPAV scheme is compared with ECPP and CAS schemes which have low verification cost in comparison with BLS, GSB and KPSD schemes as tabulated in Table 5.2.

Table 5.2 Comparisons of total computational delay of various schemes

Method	T_{gen}^{sender}	$T_{verify}^{receiver}$
ECPP	$13T_m + 6T_p$	$3T_p + 11T_m + T_h$
CAS	$2T_p + T_h + 4T_m$	$5T_p + 2T_h$
CPAV	$4T_{ep-1} + T_{ep-2} + T_h$	$2T_p + 2T_m + T_h$

From Figure 5.2, it can be seen that, the TCC of the proposed CPAV is superior to that of both ECPP and CAS schemes. Let's consider r be the authentication coefficient to indicate the ratio of vehicles which are authenticated within a communication range.

$$r = \frac{N_{authenticated}}{N_{total}}$$

Where $N_{authenticated}$ denotes the number of authenticated vehicles and N_{total} is the total number of vehicles within a communication region. In this way, the lower value of TCC increases $N_{authenticated}$ in a region, which will increase the ratio r.

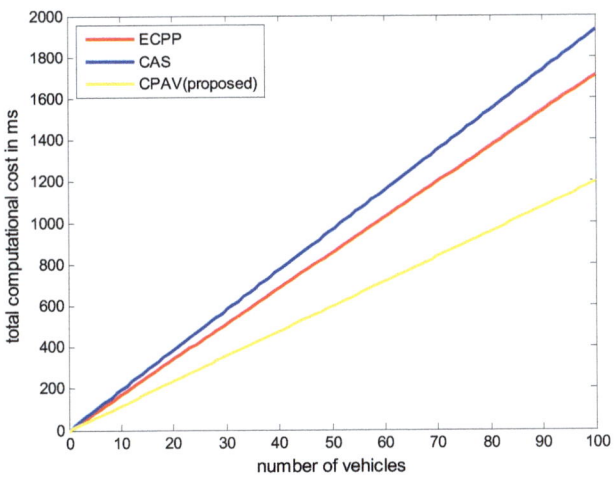

Figure 5.2. Comparisons of total computational delay

5.7   CONCLUSIONS

In this research work, a new CPAV authentication scheme is proposed for providing secure vehicular communication in VANETs. In the proposed CPAV authentication scheme, an RSU can effectively authenticate vehicles in an anonymous manner before providing safety related messages to vehicles. The CPAV authentication scheme is not only providing the anonymous authentication with low certificate and signature verification cost which is essentially required in the VANET applications, but also provide an efficient conditional privacy tracking mechanism to reveal the real identity of the malicious vehicle for enhancing the efficiency of the VANET system. The proposed CPAV authentication scheme also provides better efficiency in terms of fast verification on certificates and signatures than the previously reported schemes BLS, ECPP, CAS, GSB and KPSD.

CHAPTER 6

EFFICIENT ANONYMOUS AUTHENTICATION OF AN RSU

6.1 INTRODUCTION

Unless proper security measures are taken, the VANET users may potentially vulnerable to a number of attacks, namely, bogus information attack, impersonation attack and RSU replication attack, etc. Authentication is considered to be the first line of defence against malicious vehicles and messages. If authentication is not given, a malicious vehicle may impersonate as an RSU and sends false Location Based Safety Information (LBSI) messages to other vehicle users. If message integrity is not provided, a malicious vehicle could change the content of a message that is sent by the legitimate RSU. By doing so, the legitimate RSU would be made responsible for the damage caused. Moreover, if an anonymized vehicle or an RSU in the VANET system turns malicious, then its privacy should be revoked by the Trusted Authority (TA) and revealed to other vehicles, and then it can no longer be available in the VANET system. Thus, this revocation scheme has been considered as very essential to continue other users honest in the VANET system.

The traditional symmetric key based point-to-point source authentication mechanism is not suitable for secure source authentication in the case of multicast communications in VANETs. The symmetric key based point-to-point source authentication mechanism schemes suffer from a common drawback, that is, any receiver can impersonate the sender and then forge the data because of the shared secret key between the sender and the receiver. Moreover, the rapid change in VANET network topology due to the high-speed mobility of vehicles limits the communication time between vehicles and RSUs.

Therefore, there is a need for an effective anonymous authentication scheme that has properties like easy revocation and low computational cost. In order to facilitate this, an efficient anonymous authentication scheme with conditional privacy preservation for an RSU is proposed in this research work. The construction of this research work is based on bilinear pairing (Blake et al. 2006). In this research work, the TA generates the anonymous certificates for RSUs to protect its privacy. In the event of any conflict, the TA has the ability to successfully revoke the anonymity of a misbehaving RSU to disclose its real identity. Then the revoked identity is placed in the identity revocation list (IRL) which is maintained by the TA.

6.2 ANONYMOUS AUTHENTICATION

In this strategy, the digital signature technique is used to sign each message sent by the RSUs to the OBUs. Thus, any recipient can validate the received message and confirm the integrity and authenticity of the messages. In order to confirm the integrity and authenticity of the messages, this research work uses bilinear pairing technique, which is considered as the basis of this scheme.

6.2.1 System Initialization

From the bilinear parameters (G_1, G_2, G_T, e, q), the TA generates the system parameters as follows. The TA first selects two random numbers

$a, b \in Z_q^*$ as the master secret keys and computes $A_1 = g_1^a$ and $B_1 = g_1^b$. The TA also selects a secure cryptographic hash function $H: \{0,1\}^* \rightarrow Z_q^*$. Finally, the TA publishes the system parameters as $param = (q, e, g_1, g_2, G_1, G_2, G_T, A_1, B_1, H)$.

6.2.2 Anonymous Authentication of an RSU

Each authenticated vehicle is required to authenticate the RSU, before communicating with it. Because, each RSU provides the location based safety information (LBSI) to all authenticated vehicles when they are entered into its region. By doing this, each RSU provides the knowledge to vehicle users about the obstacles within its coverage area (Golestan et al. 2016). Table 6.1. shows some typical examples for LBSIs which are broadcasted to authenticated vehicle by RSUs.

a) **Registration and key generation:** In the registration process, Each RSU is required to submit the location information in which they are located aside the roads to the TA. After that, the TA first selects a random number $R_i \in Z_q^*$ and computes $L_i = B_1^{R_i}$. Then, the TA generates the RSU identity RSU_i and stores (RSU_i, L_i^b) in its tracking list.

b) **Anonymous certificate generation:** The TA randomly selects $\omega, \phi_1, \phi_2 \in Z_q^*$ and computes $\theta_U, \theta_V, h, h_1, h_2$.

$$\theta_U = A_1^{\phi_1} \cdot g_1^\omega, \quad \theta_V = L_i \cdot A_1^{\phi_2}, \quad h = (\omega + \phi_1) \bmod q$$

$$h_1 = \theta_V^{\phi_1 + \phi_2}$$

$$h_2 = \theta_U^{\phi_1} \cdot \theta_V^{\phi_2 - \phi_1}$$

Then, the TA computes the challenger $c_{irsu} = H(A_1 \parallel B_1 \parallel \theta_U \parallel \theta_V \parallel h_1 \parallel h_2)$ as well as $h_\alpha, h_\beta, h_\gamma$ where

$$h_\alpha = \frac{\theta_V^h}{\theta_V^\omega}$$

$$h_\beta = \theta_V^{\phi_2}$$

$$h_\gamma = \theta_U^{\phi_1}$$

Then, the TA generates an anonymous certificate for an RSU as $cert_{irsu} = \{\theta_U \parallel \theta_V \parallel c_{irsu} \parallel h_\alpha \parallel h_\beta \parallel h_\gamma\}$. After generating $cert_{irsu}$, the TA issues it to a particular RSU. Since the certificate carries no information about the RSU's real identity, it is considered as "anonymous".

c) **Signature generation:** In order to authenticate and preserve the integrity of a message $LBSI$, the RSU generates a signature by performing two simple operations:

1. The RSU first selects some random numbers $x_1, x_2, \ldots x_l \in Z_n^*$ as the temporary short time keys and computes the corresponding temporary short time public keys $Y_j = g_2^{x_j}$ for $j = 1, 2, \ldots, l$.

2. Using the temporary short time private keys x_j, the RSU generates the signature as $sig_{irsu} = g_1^{\frac{1}{x_j + H(LBSI)}}$ and broadcast $msg = (LBSI \parallel sig_{irsu} \parallel Y_j \parallel cert_{irsu})$ to all the vehicles.

Table 6.1 Location based safety information (LBSI)

Information	Range
Petrol station	20 m
Speed breaker	25 m
Traffic signal	50 m
Curve speed warning	71 m
School zone	114 m
Road intersections	170 m
Accident zone	200 m

d) Verification: Upon receiving the $msg = (LBSI \parallel sig_{irsu} \parallel Y_j \parallel cert_{irsu})$, the receiving vehicles can authenticate an RSU as follows.

1. **Certificate Verification**: To ensure the legitimacy of an RSU, the receivers first compute

$$h_1' = h_\alpha \cdot h_\beta$$

$$h_2' = \frac{h_\beta \cdot h_\gamma}{h_\alpha}$$

and then compute the challenger $c_{irsu}' = H(A_1 \parallel B_1 \parallel \theta_U \parallel \theta_V \parallel h_1' \parallel h_2')$ and checks whether $c_{irsu} = c_{irsu}'$. If it is satisfied, the certificate $cert_{irsu}$ is accepted by the receiver; otherwise it will be rejected.

Proof of correctness:

➤ $\quad h_1' = h_\alpha \cdot h_\beta$

$$= \theta_V^{h} \cdot \theta_V^{\phi_2}$$

$$= \frac{\theta_V^{\omega+\phi_1}}{\theta_V^{\omega}} \cdot \theta_V^{\phi_2}$$

$$= \theta_V^{\omega+\phi_1-\omega+\phi_2}$$

$$= \theta_V^{\phi_1+\phi_2} = h_1$$

➤ $\quad h_2' = \frac{h_\beta \cdot h_\gamma}{h_\alpha}$

$$= \frac{\theta_V^{\phi_2} \cdot \theta_U^{\phi_1}}{\theta_V^h} \cdot \theta_V^{\omega}$$

$$= \theta_V^{\phi_2+\omega-\phi_1-\omega} \cdot \theta_U^{\phi_1}$$

$$= \theta_V^{\phi_2-\phi_1} \cdot \theta_U^{\phi_1} = h_2$$

2. **Signature verification:** After verifying the certificate, the receiver then verifies the integrity of *LBSI* by checking the following condition

$$e(sig_{irsu}, Y_j \cdot g_2^{H(LBSI)}) = e(g_1, g_2)$$

If the condition is satisfied, the *LBSI* will be accepted by the receiving OBUs. Otherwise, it will be rejected.

Proof of correctness:

$$e(sig_{irsu}, Y_j \cdot g_2^{H(LBSI)}) = e(g_1^{\frac{1}{x_j+H(LBSI)}}, g_2^{x_j} \cdot g_2^{H(LBSI)})$$

$$= e(g_1^{\frac{1}{x_j+H(LBSI)}}, g_2^{x_j+H(LBSI)})$$

$$= e(g_1, g_2)^{\frac{1}{x_j+H(LBSI)} \cdot x_j + H(LBSI)}$$

(By bilinear property)

$$= e(g_1, g_2)$$

e) **Conditional tracking:** In case of compromise, the TA get the anonymous certificate of a particular RSU $cert_{irsu} = \{\theta_U \parallel \theta_V \parallel c_{irsu} \parallel h_\alpha \parallel h_\beta \parallel h_\gamma\}$ and computes

$$\frac{\theta_V{}^b}{B_1{}^{a\phi_2}} = \frac{(L_i \cdot A_1^{\phi_2})^b}{(g_1^b)^{a\phi_2}} = \frac{L_i^b \cdot A_1^{\phi_2 b}}{(g_1^a)^{b\phi_2}} = \frac{L_i^b \cdot A_1^{\phi_2 b}}{A_1^{\phi_2 b}} = L_i^b$$

So, the TA can efficiently trace the real identity of the RSU RSU_i by looking up the value L_i^b in its tracking list. After tracing the real identity, the TA will revoke the privacy of an RSU to avoid further damage.

6.3 SECURITY ANALYSIS

In this section, we briefly analyse this research work with respect to unforgeable, conditional privacy preserving, anonymity and resistance to bogus message attacks.

1. **Unforgeable:** An RSU's signature on a LBSI is defined as $sig_{irsu} = g_1^{\frac{1}{x_j+H(LBSI)}}$. Since the value x_j is only known by a particular RSU, no other users can forge the signature. Hence, without knowing the short time private keys no other user can forge the anonymous certificates and signatures.

2. **Conditional Privacy Preservation:** In this research work, RSU's use anonymous certificates and signatures to protect

their real identities. However, the TA has the capability to trace the real identity of a vehicle or an RSU from its anonymous certificate. For instance, an RSU sends a message along with an anonymous certificate to other vehicles, which leads to dispute. If it is bogus, then the TA gets the anonymous certificate of that message and maps the anonymous certificate with its tracking list. From the mapping, the TA can trace the real identity of an RSU effectively. After getting its real identity, the privacy of the vehicle user is revoked by the TA.

3. **Anonymity:** Even a valid signature sig_{irsu} and certificate $Cert_{irsu}$ are attached to each message, it is computationally hard to identify the actual signer of the message. Hence, the attacker gets zero knowledge about the signer from sig_{irsu} and $Cert_{irsu}$.

4. **Resistance to bogus message attacks:** Since each entity in the VANET checks the correctness of the received message with respect to signature attached to it, a bogus message will not pass the correctness test, and it will be finally dropped by the entities. Therefore, this research work is resistant to bogus message attacks.

6.4 PERFORMANCE ANALYSIS

This section evaluate the performance of this research work in terms of RSU serving capability for providing LBSI messages to all authenticated vehicles within its coverage region without any message loss.

6.4.1 RSU Serving Capability

When a vehicle enters into the coverage range of an RSU, the RSU first authenticates the vehicle and then the vehicle authenticates the RSU. Once the authentication process is completed, the RSU sends *LBSI* messages to the vehicle. Let p be the probability for each RSU to issue *LBSI* messages to the vehicles, and N be a random variable denoting the number of vehicles getting exactly $nLBSI$ messages among the total of d vehicles. Then, N follows the Binomial distribution with parameters d and p and we have

$$f(n;d,p) = P(N = n) = \binom{d}{n} p^n (1-p)^{d-n}$$

for $n = 0, 1, 2, \ldots, d$, where

$$\binom{d}{n} = \frac{d!}{n!\,(d-n)!}$$

is the binomial coefficient, (p^n) denotes exactly $nLBSI$ messages to successfully reach the vehicle and $p^{(d-n)}$ denotes failures. However, $nLBSI$ messages can successfully reach the vehicles, when the vehicle is anywhere in the region of RSU among the total number of d vehicles. In order to calculate the RSU serving capability, we first estimate the time required for an RSU to generate the certificate and signatures for $nLBSI$ messages. Let T_{gen} denotes the time required for an RSU to generate the certificate and signature for $nLBSI$ messages. In EAAP scheme,

$$T_{gen} = 6nT_{ep-1} + nT_p + 4nT_m$$

Therefore, the time required for an RSU to generate the certificate and signature for one *LBSI* message is as follows:

$$T_{gen} = 6T_{ep-1} + T_p + 4T_m$$

Based on the execution time results, T_{gen} time can be calculated as

$$T_{gen} = 6 \times 0.7 + 2.7 + 4 \times 0.001 = 6.904 \, ms$$

Let s denotes the average speed of a vehicle that varies from 5 m/s ~ 10 m/s (or 18 km/hr ~ 36 km/hr). Let r denotes the coverage range of an RSU which is considered as 300 m and d denotes the density of vehicles varies from 200 to 400 for a city road highway. Based on the average speed of vehicles s, the coverage range of RSU r, the probability for each RSU to issue a *LBSI* message p, and the time cost T_{gen}, the RSU serving capability R_{ser} can be calculated as

$$R_{ser} = \frac{p.T_{gen}.r}{s.d}$$

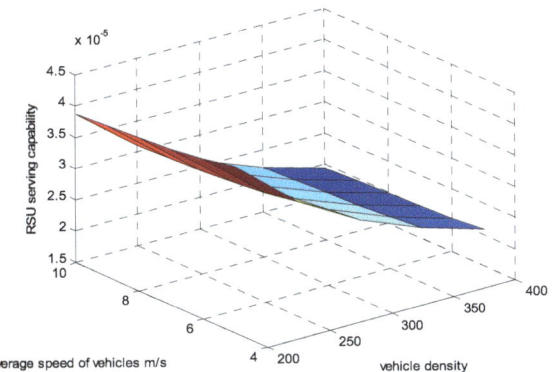

Figure 6.1 RSU serving capability of EAAP scheme for various vehicle density d and various average speed of vehicles v, when RSU range r = 300 m

Figure 6.1 shows that R_{ser} varies with vehicle density d and vehicle speed s, where $200 \leq d \leq 400$ and $5 \leq s \leq 10$. Also, we observed that the RSU can effectively generate 44 *LBSI* messages for every 300 ms and R_{ser} is inversely proportional to the RSU coverage area r, average speed of vehicle s and the vehicle density d. Therefore, in the proposed research work, the RSU can generate 44 certificates and signatures for every 300 ms but the vehicles can verify 200 certificates and signatures for every 300 ms. Therefore, it is concluded that the proposed research work provides lowest message loss when the number of vehicles within the communication range increases.

6.5 CONCLUSIONS

In this research work, a new anonymous authentication scheme for an RSU is proposed in which the vehicles can also authenticate an RSU in an anonymous manner before receiving $LBSI$ messages from them. This research work is not only providing the anonymous authentication, but also able to provide an efficient conditional privacy tracking mechanism to reveal the real identity of the malicious RSUs for enhancing the efficiency of the VANET system.

CHAPTER 7

CEKD: COMPUTATIONALLY EFFICIENT KEY DISTRIBUTION

7.1 INTRODUCTION

Several approaches have been proposed (Lu et al. 2012, Du et al. 2003, Xiaozhuo et al. 2015 and Teng & Wu 2012) to securely distribute the group key from the service provider to user side. Among the various approaches the Diffie-Hellman key exchange protocol (Joux 2000) is used for exchanging the group key because it supports the secure key distribution between two communication parties. However, the main problem in Diffie-Hellman based approaches is that the man-in-the middle attack. In this attack, and intruder secretly alters the communication between two entities who believe that they are directly communicating with each other. In order to solve this attack, a new computationally efficient key distribution scheme is developed in this research work. The main purpose of this proposed research work is to enable the TA to securely and efficiently exchange the group key to the group members in the group. Since this proposed research work depends on the hardness of computing discrete logarithms, this approach avoids man-in-the-middle attack. Moreover, this work also consumes less computational time to perform the key distribution process.

7.2 CEKD SCHEME

The proposed CEKD scheme consists of three phases namely system initialization, VANET license issuing and the proposed CEKD scheme.

7.2.1 System Initialization

The TA first picks a random number $v \in Z_q^*$ as the TA's master secret key and a random number $t \in Z_q^*$ as the TA's private key. The public key of TA is computed by performing a elliptic curve based point multiplication of t with a generator P (Johnson et al. 2001 and Koblitz 1987). Hence, $PU_{TA} = tP$ is the TA's public key. Then the TA publishes $\{q, PU_{TA}, P, G_1\}$ as the system parameters and these parameters are used for key distribution.

7.2.2 VANET License Issuing

The vehicle user first directly goes to the TA and provides their personal information (i.e., username, address, mail id, personal password, license plate number and mobile phone number) for authentication when they connect to the VANET from his/her vehicle. After providing the personal information, the TA verifies and issues the certificate CET_{V_i} to the vehicle V_i as follows.

1. TA chooses a random secret number $u_i \in Z_q^*$ for each vehicle and considers this as the private key of V_i, and computes its corresponding public key $PU_{V_i} = u_i P$.

2. TA then generates the VANET license VL_{V_i} for V_i, where $VL_{V_i} = vPU_{V_i}$.

3. Then, the TA computes a seed value $S_i = v(u_i + t)^{-1}P$ for each vehicle V_i.

4. The TA provides S_i, u_i, PU_{V_i} and VL_{V_i} to V_i in the offline mode after the successful completion of vehicle's successful registration. The values S_i and u_i are kept secret by the VANET user.

5. Next, TA keeps $(ID - V_i, VL_{V_i}, S_i, u_i)$ in its tracking list, where $ID - V_i$ is the identity of V_i assigned by the TA.

7.2.3 CEKD Scheme

a. For a secure group communication, each vehicle user V_i first selects a session key $x_i \in Z_q^*$ and computes

$$X_i = x_i(PU_{V_i} + PU_{TA})$$

$$X_i = x_i(u_i P + tP)$$

$$X_i = x_i(u_i + t)P$$

The main purpose of using this session key is that it is used to compute the X_i value which is the transferring parameter from the user side to the TA side for the identification key computation. Then, each user also computes the identification key (K_i) as follows

$$K_i = e(X_i, S_i)$$

$$K_i = e(x_i(u_i + t)P, v(u_i + t)^{-1}P)$$

$$K_i = e(P,P)^{x_i(u_i+t) \ast v(u_i+t)^{-1}}$$

$$K_i = e(P,P)^{x_i \ast v}$$

After computing the value of K_i, each user sends $\{X_i \parallel VL_{v_i}\}$ to the TA. After receiving these parameters from the vehicle user V_i, the TA verifies the VL_{v_i} from it's tracking list and then it computes the K_i value for each vehicle user V_i.

b. Then, the TA chooses a group key $gk \in Z_q^*$ and then creates a Lagrange interpolating polynomial of degree $n-1$, where n is the number of users in the group.

$$P(y) = \sum_{i=1}^{n}(gk+u_i)\left[\prod_{\substack{j=1 \\ j \neq i}}^{n} \frac{y-K_j}{K_i-K_j}\right]$$

$$= a_0 + a_1 y^1 + \cdots + a_{n-1} y^{n-1}$$

Then the TA sends $(a_0, a_1, \ldots a_{n-1})$ to all the users available in the group.

c. After receiving the coefficients $(a_0, a_1, \ldots a_{n-1})$ from the TA each user V_i uses its own K_i value to recover the group key gk as follows:

$$P(K_i) = a_0 + a_1 K_i^1 + \cdots + a_{n-1} K_i^{n-1}$$

$$= gk + u_i$$

Then, each user recovers the group key by subtracting its own private key from $gk + u_i$. This group key is used to make group communication with the group members in the group.

7.3 SECURITY ANALYSIS

In this section, the security strength of this research work is evaluated with respect to the user authentication, impersonation attack and man-in-the-middle attack.

1. **User Authentication:** The VANET license $VL_{V_i} = vPU_{V_i}$ for each vehicle user V_i is computed by the TA using each user's private key and its master key. So, it is assured that no other users compute VL_{V_i} except the TA who has both the private key of V_i and the master key. Since, each vehicle user V_i sends $\{X_i \parallel VL_{V_i}\}$ to the TA for getting the group key, the TA can authenticate V_i from VL_{V_i} before providing the group key to them.

2. **Impersonation**: Consider for example, V_i's private key and its license VL_{V_i} is revealed. An adversary U wants to pretend as the vehicle user V_i and communicates with the TA to get the group key. However, U cannot compute the identification key K_i without knowing the seed value S_i which is given by the TA to V_i during the time of user registration.

3. **Man-in-the-middle attack:** Based on the above analysis with respect to the user authentication, the proposed CEKD scheme provides secure key distribution with authentication between the communicating parties. Therefore, the CEKD scheme could withstand from the man-in-the-middle attack. In this way, it is not possible to perform man-in-middle attack in this proposed work.

7.4 PERFORMANCE ANALYSIS

In this section, the performance of the CEKD scheme is compared with the previously proposed key distribution schemes (Lu et al. 2012, Du et al. 2003, Xiaozhuo et al. 2015 and Teng & Wu 2012) in terms of computational cost. The computational cost is defined as the total time required for the vehicle user to get the group key from the TA. The computational cost of CEKD scheme is compared with many existing schemes, namely DIKE (Lu et al. 2012), ID-AGKA (Du et al. 2003), HPF-CLGKA (Xiaozhuo et al. 2015) and Teng's scheme (Teng & Wu 2012). The notations used in this approach are T_p, T_h and T_m where T_p represents the time required for performing a pairing operation, T_h represents the time required for performing a hash operation and T_m represents the time required for performing a point multiplication.

For performing the hash operation, point multiplication and pairing operation, the pairing-based cryptography (PBC) library is used in this research work. For the aforementioned operations, the Type-A curve defined in the PBC library is used with the default parameters. In order to measure the actual computation time of the proposed CEKD scheme, a 2-GHz machine with 4-GB installed memory, running Cygwin 1.7.35–15 with the gcc version 4.9.2 is used for the implementations.

All the results are analysed over 100 randomized simulation runs and then the average of the results is considered. In the simulations, the time parameters T_p, T_h and T_m are measured and it is found to be equal to 1.6 ms (milliseconds), 2.7 ms, and 0.001 ms, respectively.

Table 7.1 Computational cost of various key distribution schemes

Method	For one user	For n users
DIKE	$3T_p + 2T_h$	$(2n+1)T_p + 2nT_h$
ID-AGKA	$4T_p + 6T_m$	$4nT_p + n(n+1)T_m$
HPF-CLGKA	$9T_m + 2T_h$	$9nT_m + 2nT_h$
Teng's scheme	$2T_p + 2T_m + T_h$	$2nT_p + 2nT_m + nT_h$
CEKD (Proposed)	$2T_p + T_m$	$2nT_p + nT_m$

From Table 7.1, it can be observed that the proposed CEKD scheme requires only two pairing and one point multiplication operation to get the gk from the TA and hence CEKD takes low computational cost compared to other existing schemes.

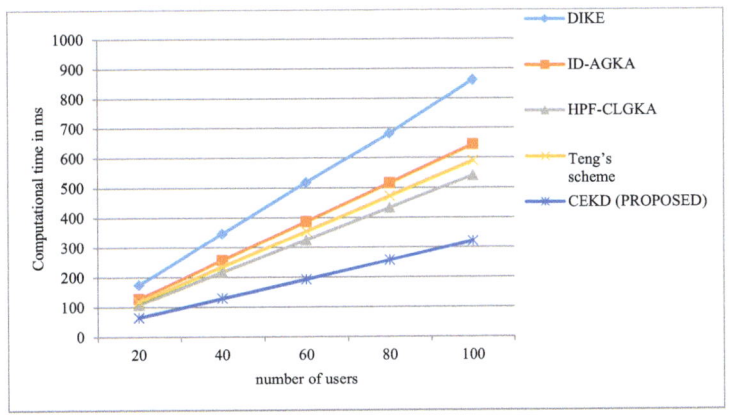

Figure 7.1 Computational cost for key distribution of various schemes

From Figure 7.1, it is very clear to understand that the CEKD scheme takes only 320 ms for a secure group key distribution from the TA to the vehicle user. However, other existing schemes take more than 500 ms for secure group key distribution.

7.5 CONCLUSIONS

In this research work, CEKD scheme is proposed for secure group key distribution in VANETs. In order to avoid the man-in-the-middle attack, the TA performs the authentication process for each vehicle user V_i before distributing the group key. The security analysis section shows that the weaknesses of previously proposed schemes can be overcome by the proposed CEKD scheme and hence the proposed work improves the security level requirements. Moreover, the performance analysis section shows that the proposed CEKD scheme takes low computational cost which makes it suitable for the VANET environment.

CHAPTER 8

CONCLUSIONS AND FUTURE WORKS

It is an unquestionable fact that every year road accidents becomes one of the major causes of death all over the world. In order to avoid these accidents, VANET has been introduced as an advanced solution. Since, VANET systems have a wide range of safety and non-safety applications, security services take a vital role to safeguard the VANET system from various kind of security attacks.

8.1 DUAL AUTHENTICATION AND DUAL KEY MANAGEMENT FOR GROUP COMMUNICATION

In this research work, a new dual authentication scheme has been proposed for improving the security of vehicles that are communicating with the VANET environment. For providing such authentication in dual mode, two components are used namely hash code and fingerprint of each communicating vehicle user. Therefore, the fingerprint authentication technique is integrated into a hash code creation method in this research work to avoid malicious users to use the secret key of any VANET users in order to participate in the VANET communication. Moreover, to avoid malicious users from spoofing the authentication code issued for any VANET users and sending erroneous messages to other vehicles a new dual key management scheme has been introduced in this research paper. The dual key management scheme implemented in this research work is computationally efficient that supports secure data transmission from TA to PUs and Pus to SUs based on two different group keys, one for PUs and another one for SUs for further improving the security among different classes of vehicles. Moreover, the proposed algorithm also takes single broadcast messages from TA to inform the group members in order to recover the updated group key.

8.2 CPAV: COMPUTATIONALLY EFFICIENT PRIVACY PRESERVING ANONYMOUS AUTHENTICATION

In this research work, a new CPAV authentication scheme have been proposed for secure vehicular communication in VANETs. In the proposed CPAV authentication scheme, an RSU can effectively authenticate vehicles in an anonymous manner before providing safety related messages to vehicles. The CPAV authentication scheme is not only providing the anonymous authentication with low certificate and signature verification cost which is essentially required in the VANET applications. It also provides an efficient conditional privacy, tracking mechanism to reveal the real identity of the malicious vehicle for enhancing the efficiency of the VANET system. The proposed CPAV authentication scheme also provides better efficiency in terms of fast verification on certificates and signatures than the previously reported schemes BLS, ECPP, CAS, GSB and KPSD.

8.3 EFFICIENT ANONYMOUS AUTHENTICATION OF AN RSU

In this research work, an efficient anonymous authentication of an RSU have been proposed for secure vehicular communication in VANETs. In the proposed anonymous authentication scheme, vehicles can also authenticate an RSU in an anonymous manner before receiving *LBSI* messages from RSUs. Moreover, to avoid communication with the compromised RSUs an efficient conditional privacy tracking mechanism have been proposed to reveal the real identity of the malicious RSUs for enhancing the efficiency of the VANET system.

8.4 CEKD: COMPUTATIONALLY EFFICIENT KEY DISTRIBUTION

In this research work, CEKD scheme have been proposed for secure group key distribution in VANETs. The TA performs the authentication process for each vehicle user before distributing the group key in order to avoid the man-in-the-middle attack. The security analysis and performance analyses sections show that the proposed scheme is both secure in terms of man in the middle attack and efficient in terms of computational cost.

8.5 FUTURE WORKS

The future development of this work is to develop an efficient batch authentication scheme to anonymously authenticate more number of vehicles or messages simultaneously to alleviate the authentication burden greatly. Moreover, the future work is extended to develop a new location based privacy enhanced techniques under a stronger threat model.

REFERENCES

1. Akila, M & Iswarya, T 2011, 'An efficient data replication method for data access applications in VANETs', Proceedings of the IEEE international conference on electronics, communication and computing technologies, pp.17-22.

2. Al-kahtani, MS 2012, 'Survey on security attacks in vehicular ad hoc networks (VANETs)', Proceedings of the 6th international conference on signal processing and communication systems, pp. 1–9.

3. Ameneh, D & Ghaffar, PRA 2013, 'Detection of malicious vehicles (DMV) through monitoring in vehicular ad-hoc networks', Springer Multimedia Tools and Applications, vol. 66, no. 2, pp. 325-338.

4. Biswas, S & Misic, J 2013, 'A cross-layer approach to privacy-preserving authentication in WAVE-enabled VANETs', IEEE Transactions on Vehicular Technology, vol. 62, no.5, pp. 2182–2192.

5. Blake, I, Murty, V & Xu, G 2006, 'Refinements of Miller's algorithm for computing the Weil/Tate pairing', Journal of Algorithms, vol. 58, no. 2, pp. 134–149.

6. Blum, J & Eskandarian, A 2004, 'The threat of intelligent collisions', IT Professional, vol.6, no.1 pp. 24–29.

7. Boneh, D, Gentry, C, Lynn, B & Shacham, H 2003, 'Aggregate and verifiably encrypted signatures from bilinear maps', Proceedings of Advances in Cryptology. pp. 416–432.

8. Boneh, D, Lynn, B & Shacham, H 2004, 'Short signatures from the Weil Pairing', Journal of Cryptology, vol. 17, no. 4, pp. 297–319.

9. Busanelli, S, Ferrai, G & Veltri, L 2011, 'Short-lived key management for secure communications in VANETs', Proceedings of the IEEE international conference on ITS telecommunication, pp. 613-618.

10. Cheng, X, Yang, L & Shen, X 2015, 'D2D for intelligent transportation systems : a feasibility study', IEEE Transactions on intelligent transportation systems, vol. 16, no. 4, pp. 1784-1793.

11. Chima, TW, Yiu, SM, Hui, LCK & Li, VOK 2011, 'SPECS: secure and privacy enhancing communications schemes for VANETs', Journal of Ad Hoc Networks , vol. 9, no.2, pp. 189-203.

12. Choi, J & Jung, S 2009, 'A security framework with strong non-repudiation and privacy in VANETs', Proceedings of the 6th IEEE conference on consumer communications and networking conference, pp.1-5.

13. Cygwin 2016, Linux Environment Emulator for Windows. Available from: http://www.cygwin.com. [11 April 2016]

14. Das, ML, Saxena, A, Gulati, VP & Phatak, DB 2006, 'A novel remote user authentication scheme using bilinear pairings', Computers and Security, vol. 25,no.3, pp. 184-189.

15. Dhamgaye, A & Chavhan, N 2013, 'Survey on security challenges in VANET', International Journal of Computer Science and Network Security, vol. 2, no. 1, pp. 88–96, 2013.

16. Douceur, JR 2002, 'The sybil attack', Proceedings of the first international workshop on peer-to-peer systems, pp. 251–260.

17. Du, X, Wang, Y, Ge, J & Wang ,Y 2003, 'An ID-based Authenticated Two Round Multi-Party Key Agreement', Proceedings of the international association for crypto research, pp 247-254.

18. Gamage, C, Gras, B, Crispo, B & Tanenbaum, AS 2006, 'An identity based ring signature scheme with enhanced privacy', Proceedings of the IEEE secure communication and workshops, pp. 1–5.

19. Ghosh, M, Varghese, A, Kherani, AA & Gupta, A 2009,'Distributed misbehavior detection in VANETs', Proceedings of the IEEE wireless communication and networking conference, pp.1-6.

20. Golestan, K, Khaleghi, B, Karray, F & Kamel, MS 2016, 'Attention assist: a high-level information fusion framework for situation and threat assessment in vehicular ad hoc networks', IEEE Transactions on Intelligent Transport Systems, vol. 17, no. 5, pp. 1271 – 1285.

21. Gong, Z, Long, Y, Hong, X & Chen, K 2007, 'Two certificate less aggregate signatures from bilinear maps', Proceedings of the 8^{th}ACIS international conference on software engineering, artificial

intelligence, networking, and parallel/ distributed computing, pp.188–193.

22. Grover, J, Gaur, MS, Laxmi V & Prajapati NK 2011, 'A sybil attack detection approach using neighboring vehicles in VANET', Proceedings of the 4th international conference on security of information and networks, pp.151-158.

23. Guo, J, Baugh, JP & Wang, S 2007, 'A group signature based secure and privacy-preserving vehicular communication framework', Proceedings of the mobile networking for vehicular environments workshop in conjunction with IEEE INFOCOM, pp.103-108.

24. Hao, Y, Cheng, Y & Ren, K 2008, 'Distributed key management with protection against RSU compromise in group signature based VANET', Proceeding of the IEEE global telecommunications conference, pp.1-5.

25. Hao, Y, Cheng, Y, Zhou, C & Song, W 2011, 'A distributed key management framework with cooperative message authentication in VANETs', IEEE Journal on Selected Areas in Communications, vol. 29, no. 3, pp. 616–629.

26. Hawi, FA, Yeun, CY & Qutayti, MA 2009, 'Security challenges for emerging VANETs', Proceedings of the 4thinternational conference on information technology, pp. 3–5.

27. He, L & Zhu, WT 2012, 'Mitigating dos attacks against signature-based authentication in VANETs', Proceedings of the 2012 IEEE international conference on computer science and automation engineering, pp. 261–265.

28. Huang, JL, Yeh, LY & Chien, HY 2011, 'ABAKA: An anonymous batch authenticated and key agreement scheme for value-added services in vehicular ad hoc networks', IEEE Transactions on Vehicular Technology, vol. 60, no. 1,pp. 248–262.

29. Huang, D, Misra, S, Verma, M & Xue, G 2011, 'PACP: An efficient pseudonymous authentication-based conditional privacy protocol for VANETs,' IEEE Transactions on Intelligent Transportation Systems, vol. 12, no. 3, pp. 736–746.

30. Jakubiak, J & Koucheryavy, Y 2008, 'State of the art and research challenges for VANETs', Proceedings of the 5th IEEE consumer communications and networking conference, pp. 912-916.

31. Jia, X, Yuan, X, Meng, L & Wang, L 2013, 'EPAS: Efficient privacy-preserving authentication scheme for VANETs-based emergency communication', Journal of Software, vol. 8, no.8, pp. 1914-1922.

32. Jiang, D & Delgrossi, L 2008, 'IEEE 802.11p: towards an international standard for wireless access in vehicular environments', Proceedings of the IEEE 68th vehicular technology conference, pp. 2036–2040.

33. Johnson, D, Menezes, A & Vanstone, S 2001, 'The elliptic curve digital signature algorithm (ECDSA)', International Journal of Information Security, vol. 1, no. 1, pp. 36–63.

34. Joux, A 2000, 'A one round protocol for tripartite Diffie-Hellman', Proceedings of the 4th international symposium on algorithmic number theory, pp.385-393.

35. Kassim, M, Rahman, RA & Mustapha, R 2011, 'Mobile ad hoc network (MANET) routing protocols comparison for wireless sensor network', Proceedings of the IEEE international conference on system engineering and technology, pp. 148-152.

36. Kitani, T, Shinkawa,T, Shibata, N, Yasumoto, K, Ito, M & Higashino, T 2008, 'Efficient VANET-based traffic information sharing using buses on regular routes', Proceedings of the IEEE vehicular technology conference, pp.3031-3036.

37. Koblitz, N 1987, 'Elliptic curve cryptosystems', Mathematics of Computation, vol. 48, no. 177, pp. 203–209.

38. Li, J, Lu, H & Guizani, M 2014, 'ACPN: A novel authentication framework with conditional privacy-preservation and non-repudiation for VANETs', IEEE Transactions on Parallel and Distributed Systems, vol.26, no.4, pp-938 – 948.

39. Lin, X & Li, X 2013,'Efficient cooperative message authentication in vehicular ad hoc networks', IEEE Transactions on Vehicular Technology, vol.62, no.7, pp. 3339–3348.

40. Lin, X, Lu, R, Liang, X & Shen X 2011, 'STAP: A social-tier-assisted packet forwarding protocol for achieving receiver-location privacy preservation in VANETs', Proceedings of the 30thIEEE conference on computer communications, pp. 2147–2155.

41. Lin, X, Sun, X, Ho, PH & Shen, X 2007, 'GSIS: A secure and privacy preserving protocol for vehicular communication', IEEE Transactions on Vehicular Technology, vol. 56, no. 6, pp. 3442–3456.

42. Lin, X, Lu, R, Zhang, C, Zhu, H, Ho, PH & Shen, X 2008, 'Security in vehicular ad hoc networks', IEEE Communications Magazine, vol. 46, no. 4, pp. 88–95.

43. Lo, NW & Tsai, HC 2007, 'Illusion attack on VANET applications-A message plausibility problem', Proceedings of the IEEE globecom workshops, pp.1-8.

44. Lu, R, Lin, X, Liang, X & Shen, X 2012, 'A dynamic Privacy-Preserving Key Management Scheme for Location-Based Services in VANET', IEEE Transactions on Intelligent Transportation Systems, vol.13, no.1, pp. 127-139.

45. Lu, R, Lin, X & Luan, TH 2009, 'Pseudonym changing at social spots: an effective strategy for location privacy in VANET', IEEE Transaction on Vehicular Technology, vol. 61, no. 1, pp. 86–96.

46. Lu, R, Lin, X, Luan, TH, Liang, X & Shen, X 2011, 'Anonymity analysis on social spot-based pseudonym changing for location privacy in VANETs', Proceedings of the IEEE International conference on communications, pp.1-5.

47. Lu, R, Lin, X, Zhu, H, Ho, P & Shen, X 2008, 'ECPP: Efficient conditional privacy preservation protocol for secure vehicular communications', Proceedings of the IEEE The 27th conference on computer communications, pp. 1229–1237.

48. Lv, X, Li, H & Wang, B 2012, 'Group key agreement for secure group communication in dynamic peer systems', Journal of Parallel and Distributed Computing, vol. 72, no. 10, pp. 1195–1200.

49. Malla, AM & Sahu, RK 2013, 'Security attacks with an effective solution for dos attacks in VANET', International Journal of Computer Applications, vol. 66, no.22, pp.45-49.

50. Marti, S, Giuli, TJ, Lai, K & Baker, M 2000, 'Mitigating routing misbehaviour in mobile ad hoc networks', Proceedings of the 6th annual international conference on mobile computing and networking, pp. 255–265.

51. Matusiewicz, K, Pieprzyk, J, Pramstaller, N, Rechberger, C & Rijmen, V 2005, 'Analysis of simplified variants of SHA-256', Proceedings of the western european workshop on research in cryptology, pp. 1–12.

52. Mershad, K & Artail, H 2013,'A framework for secure and efficient data acquisition in vehicular Ad Hoc networks', IEEE Transactions on Vehicular Technology, vol. 62, no. 2, pp. 536–551.

53. Misra, S, Bhattarai, K & Xue, G 2011, 'BAMBi: blackhole attacks mitigation with multiple base stations in wireless sensor networks', Proceedings of the IEEE international conference on communications, pp.1-5.

54. Morgan, YL 2010,' Notes on DSRC & WAVE standards suite: Its architecture, design and characteristics', IEEE Communication Surveys and Tutorials, vol.12, no.4, pp.504-518.

55. Mpitziopoulos, A, Gavalas, D, Konstantopoulos, C & Pantziou ,G 2009,'A survey on jamming attacks and countermeasures in WSNs', IEEE Communications Surveys and Tutorials,vol.11,no.4, pp. 42–56.

56. Naranjo, JAM , Ramos, JAL & Casado, LG 2012, 'A suite of algorithms for key distribution and authentication in centralized secure multicast environments', Journal of Computational and Applied Mathematics, vol. 236, no. 12, pp. 3042–3051.

57. Oh, H, Yae, C, Ahn, D & Cho, H 1999, '5.8 GHz DSRC packet communication system for ITS Services', Proceedings of the IEEE VTS 50th vehicular technology conference, pp. 2223-2227.

58. Okamoto, J & Ishihara, S 2010, 'Distributing location dependent data in VANETs by guiding data traffic to high vehicle density areas', Proceedings of the IEEE vehicular networking conference, pp.189-196.

59. Pairing-Based Cryptography [PBC] Library. Available from : http://crypto.stanford.edu/pbc/ [05 July 2005]

60. Papadimitratos, P, Gligor, V & Hubaux, JP 2006, 'Securing vehicular communications-assumptions, requirements, and principles', Proceedings of the 4th workshop on embedded security in cars, pp. 5–14.

61. Park, S & Lee, S 2012, 'Improving data accessibility in vehicle ad hoc network', International Journal of Smart Home, vol.6, no.4, pp.169-176.

62. Park, S, Aslam, B, Turgut, D & Zou, CC 2009, 'Defense against sybil attack in vehicular ad hoc network based on roadside unit support', Proceedings of the IEEE military communications conference, pp.1-7.

63. Parno, B & Perrig, A 2005, 'Challenges in securing vehicular networks', Proceedings of the workshop on hot topics in networks, pp. 1–6.

64. Raj, PN & Swadas, PB 2009, 'DPRAODV: A dyanamic learning system against blackhole attack in AODV based MANET', International Journal of Computer Science Issues, vol. 7, no.4, pp.54.

65. Rawat, A, Sharma, S & Sudhil, R 2012, 'VANET: Security attacks and its possible solutions', International Journal of Information and Operations Management, vol. 3,no.1, pp. 301–304.

66. Raya, M & Hubaux, J 2007, 'Securing vehicular ad-hoc networks', Journal of Computer Security, vol. 15, no. 1, pp. 39–68.

67. Raya, M & Hubaux, JP 2004, 'DOMINO: A system to detect greedy behavior in IEEE 802.11 hotspots', Proceedings of the 2nd international conference on mobile systems, applications and services, pp.84-97.

68. Raya, M & Hubaux, JP 2005, 'Security aspects of inter-vehicle communications', Proceedings of the 5th swiss transport research conference, pp.1-15.

69. Raya, R, Papadimitratos, P & Hubaux, JP 2006, 'Securing vehicular communications', IEEE wireless communications, vol. 13, no. 5, pp. 8–15.

70. Robinson, CL, Caveney, D, Caminiti, L, Baliga, G, Laberteaux & Kumar, PR 2007, 'Efficient message composition and coding for cooperative vehicular safety applications', IEEE Transactions on Vehicular Technology, 2007, vol.56, no.6, pp. 3244–3255.

71. Roselinmary, S, Maheshwari, M & Thamaraiselvan, M 2013, 'Early detection of DOS attacks in VANET using attacked packet detection algorithm (APDA)', Proceedings of the IEEE information communication and embedded systems, pp.237-240.

72. Shen, W, Liu, L & Cao, X 2013, 'Cooperative message authentication in vehicular cyber-physical systems', IEEE Transactions on Emerging Topics in Computing vol. 1, no. 1, pp. 84–97.

73. Shen, X, Cheng, X, Yang, L, Zhang, R & Jiao, B 2014, 'Data dissemination in VANETs: A scheduling approach', IEEE Transactions on Intelligent Transportation Systems, vol. 15, no. 5, pp. 411–416.

74. Shim, KA 2012, 'CPAS: An efficient conditional privacy preserving authentication scheme for vehicular sensor networks', IEEE Transactions on Vehicular Technology, vol.61, no.4, pp.1874-1883.

75. Sun, J, Zhang, C, Zhang, Y & Fang, Y 2010, 'An identity based security system for user privacy in vehicular a hoc networks', IEEE Transactions On Parallel And Distributed Systems, vol. 21, no.9, pp.1227 – 1239.

76. Sun, X, Lin, X & Ho, PH 2007,'Secure vehicular communications based on group signature and ID-based signature scheme', Proceedings of the IEEE international conference on communications, pp.1539–1545.

77. Sun, Y, Lu, R, Lin, X, Shen, X & Su, J 2010, 'An efficient pseudonymous authentication scheme with strong privacy preservation for vehicular communications', IEEE Transactions on Vehicular Technology, vol.59, no.7, pp. 3589-3603.

78. Syamsuddin, I, Dillon, T, Chang, E & Han, S 2008, 'A survey of RFID authentication protocols based on hash chain method', Proceedings of 3rd institute of communication, culture, information and technology, vol. 2, pp. 559–564.

79. Tan, Z 2010, 'A privacy-preserving mutual authentication protocol for vehicle Ad Hoc networks', Journal of Convergence Information Technology, vol. 5, no. 7, pp. 180–186.

80. Teng, JK & Wu, CK 2012,' A provable authenticated certificate less group key agreement with constant rounds', Journal of Communications and Networks, vol.14, no.10, pp. 104-110.

81. Veltri, L, Cirani, S, Busanelli, S & Ferrari, G 2013,'A novel batch based group key management protocol applied to the Internet of things', Ad Hoc Networks, vol. 11, no. 8, pp. 2724–2737.

82. Vighnesh, NV, Kavita, N, Shalini, R & Sampalli, S 2011,'A novel sender authentication scheme based on hash chain for vehicular ad-hoc networks', Proceedings of the IEEE Symposium on wireless technology and applications, pp. 96–101.

83. Vijayakumar, P, Bose, S & Kannan, A 2013, 'Centralized key distribution protocol using the greatest common divisor method', Computers and Mathematics with Applications, vol. 65, no. 9, pp. 1360–1368.

84. Wasef, A & Shen, X 2009, 'MAAC: Message authentication acceleration protocol for vehicular ad hoc networks', Proceedings of the IEEE global telecommunications conference, pp.1-6.

85. Wasef, A, Jiang, Y & Shen, X 2008, 'ECMV: Efficient certificate management scheme for vehicular networks', Proceedings of the IEEE global telecommunications conference, pp. 1–5.

86. Wen, H, Huang, PY, Dyer, J, Archinal, A & Fagan, J 2011, 'Countermeasures for GPS signal spoofing', Proceedings of the 18th international technical meeting of the satellite division of the institute of navigation, pp.1285-1290.

87. Wischhof, L, Ebner, A & Rohling, H 2005, 'Information dissemination in self-organizing inter vehicle networks', IEEE Transactions on Intelligent Transportation Systems vol. 6, no. 1, pp. 90–101.

88. Wong, C, Gouda, M & Lam, S 2000, 'Secure group communications using key graphs', IEEE/ACM Transactions on Networking, vol. 8, no. 1, pp. 16–30.

89. Xiao, B, Yu, B & Gao, C 2006, 'Detection and localization of sybil nodes in VANETs', Proceedings of the workshop on dependability issues in wireless ad hoc networks and sensor networks, pp. 1-8.

90. Xiaozhuo, G, Zhenjiang, C & Yongming, W 2015,' How to Get Group Key Efficiently in Mobile Ad Hoc Networks?', Proceedings of the military communications conference, pp. 1009-1014.

91. Zaharuddin, MHM, Rahman, RA & Kassim, M 2010,'Technical comparison analysis of encryption algorithm on site-to-site IPSec VPN' , Proceedings of the international conference on computer applications and industrial electronics, pp. 641-645.

92. Zeadally, S, Hunt, R, Chen, YS, Irwin A & Hassam, A 2012, 'Vehicular ad hoc networks (VANETs): status, results, and challenges', Telecommunication Systems, vol. 50, no.4, pp. 217–241.

93. Zhang, C, Lin, X, Lu, R & Ho, PH 2008, 'RAISE: An efficient RSU-aided message authentication scheme in vehicular communication networks', Proceedings of the IEEE international conference on communications, pp. 1451–1457.

94. Zhang, C, Lu, R, Lin, X, Ho, PH & Shen, X 2008, 'An efficient identity based batch verification scheme for vehicular sensor networks', Proceedings of the IEEE INFOCOM, pp. 246-250.

95. Zhang, C, Lin, X, Lu, R, Ho, PH & Shen, X 2008, 'An efficient message authentication scheme for vehicular communications, IEEE Transaction Vehicular Technology, vol.57, no.6, pp. 3357-3368.

96. Zhang, R, Cheng, X, Yang, L, Shen, X & Jiao, B 2015, 'A novel centralized TDMA based scheduling protocol for vehicular networks', IEEE Transactions on Intelligent Transportation Systems, vol. 16, no. 1, pp. 411–416.

97. Zheng, XL, Huang, CT & Matthews, M 2007, 'Chinese remainder theorem based group key management', Proceedings of the 45th ACM south east conference, pp. 266–271.

98. Zhou, J & Ou, OH 2009, 'Key tree and Chinese remainder theorem based group key distribution scheme', Journal of the Chinese Institute of Engineers, vol. 32, no. 7, pp. 967–974.

99. Zhu, H, Lin, X, Lu, R, Ho, P & Shen, X 2008, 'AEMA: An aggregated emergency message authentication scheme for enhancing the security of vehicular ad hoc networks', Proceedings of the IEEE International Conference on Communications, pp.1436–1440.

100. Zhu, J & Roy, S 2003, 'MAC for dedicated short range communication in intelligent transport system', IEEE Communications Magazine, vol. 41, no. 12, pp. 60-67.